VALUES OF SELECTED PHYSICAL CONSTANTS

Quantity	Symbol	Units
Avogadro's number	N_A	6.022×10^{23} molecules/mol
Boltzmann's constant	k	1.38×10^{-23} J/atom-K
		8.26×10^{-5} eV/atom-K
Electron charge	e	1.602×10^{-19} C
Electron mass	—	9.11×10^{-31} kg
Gas constant	R	8.314 J/mol-K
		1.987 cal/mol-K
Permeability of a vacuum	μ_0	1.257×10^{-6} henry/m
Planck's constant	h	6.626×10^{-34} J-s
Velocity of light in a vacuum	c	2.998×10^8 m/s

UNIT ABBREVIATIONS

atm = atmosphere
A = ampere
Å = angstrom
Btu = British thermal
    unit
C = Coulomb
°C = degrees Celsius
cal = calorie (gram)
cm = centimeter
eV = electron volt
°F = degrees
    Fahrenheit
ft = foot
g = gram

gal = gallon
hp = horsepower
hr = hour
in. = inch
J = joule
K = degrees Kelvin
kg = kilogram
l = liter
lb_f = pound force
lb_m = pound mass
m = meter
Mg = megagram
min = minute

mm = millimeter
mol = mole
MPa = megapascal
N = newton
nm = nanometer
Pa = pascal
psi = pounds per
    square inch
s = second
T = temperature
W = watt
μ = micrometer
    (micron)

SI MULTIPLE AND SUBMULTIPLE PREFIXES

Factor by Which Multiplied	Prefix	Symbol	Factor by Which Multiplied	Prefix	Symbol
10^{18}	exa	E	10^{-3}	milli	m
10^{15}	peta	P	10^{-6}	micro	μ
10^{12}	tera	T	10^{-9}	nano	n
10^{9}	giga	G	10^{-12}	pico	p
10^{6}	mega	M	10^{-15}	femto	f
10^{3}	kilo	k	10^{-18}	atto	a
10^{-2}	centi	c			

1-6-97

Thermodynamics of Materials

Volume II

MIT Series in Materials Science & Engineering Series Statement

In response to the growing economic and technological importance of polymers, ceramics, advanced metals, composites, and electronic materials, many departments concerned with materials are changing and expanding their curricula. The advent of new courses calls for the development of new textbooks that teach the principles of materials science and engineering as they apply to all classes of materials.

The MIT Series in Materials Science and Engineering is designed to fill the needs of this changing curriculum.

Based on the curriculum of the Department of Materials Science and Engineering at the Massachusetts Institute of Technology, the series will include textbooks for the undergraduate core sequence of courses on Thermodynamics, Physical Chemistry, Chemical Physics, Structures, Mechanics, and Transport Phenomena as they apply to the study of materials. More advanced texts based on this core will cover the principles and technologies of different material classes, such as ceramics, metals, polymers, and electronic materials.

The series will define the modern curriculum in materials science and engineering as the discipline changes with the demands of the future.

The MIT Series Committee

Samuel L. Allen
Yet-Ming Chiang
Merton C. Flemings
David V. Ragone
Julian Szekely
Edwin L. Thomas

Thermodynamics
of Materials

Volume II

David V. Ragone
Massachusetts Institute of Technology

John Wiley & Sons. Inc.
New York • Chichester • Brisbane • Toronto • Singapore

Acquisitions Editor	Cliff Robichaud
Marketing Manager	Susan Elbe
Senior Production Editor	Charlotte Hyland
Designer	Kevin Murphy
Manufacturing Manager	Susan Stetzer
Illustration	Sandra L. Rigby

This book was set in 10/12 Times Roman by University Graphics and printed and bound by R. R. Donnelley. The cover was printed by Phoenix Color.

Library of Congress Cataloging in Publication Data:
Ragone, David V., 1930–
      Thermodynamics of materials / David V. Ragone.
            p.   cm.
      Includes bibliographical references.
            ISBN 0-471-30885-4 (v. 1 : cloth).—ISBN 0-471-30886-2
(v. 2 :cloth)
            1. Materials—Thermal properties.   2. Thermodynamics.
      TA418.52.R34 1995
536′.7—dc20   94-25647
                                                                      CIP

Printed in the United States of America

10 9 8 7 6 5 4 3 2 1

To My Parents, Mary and Fred

Preface—Vol. II

This volume is a textbook for the second part of a two-course sequence at the undergraduate level on the physical chemistry of materials. The first volume deals with the laws of thermodynamics, property relations, equilibrium, phase diagrams, and an introduction to statistical thermodynamics.

Chapter 1 of this volume reviews the principles of macroscopic thermodynamics. Chapter 2 begins with a review of the principles of statistical thermodynamics and then treats such kinetic phenomena as evaporation from surface, mean free path of molecules in gases, diffusion, and the stress–strain behavior of elastomers. Subsequent chapters deal with the thermodynamics of defects and of interfaces. The latter part of the course deals with kinetic phenomena in the solid state, such as diffusion, nucleation, spinodal decomposition, and reaction kinetics. A chapter on nonequilibrium thermodynamics is included because it provides an introduction to several subjects of interest to students of materials science and engineering. This chapter covers the Seebeck effect (thermocouples), the Peltier effect, thermal diffusion (observed in materials exposed to temperature gradients), and electromigration, a phenomenon that is all too frequently observed in electronic microcircuits.

The title of this volume, *Thermodynamics, Volume II* may seem strange considering that half the book is devoted to kinetics. One could argue, however, that the study of equilibrium phenomena should be called thermo*statics,* because it deals with states that do not change. Thermo*dynamics,* then, would refer to the study of changing states. To avoid a long, needlessly complex title, *Thermodynamics* was chosen for both volumes.

I acknowledge, with thanks, the efforts of my colleagues, Profs. Kirk Kolenbrander and David Roylance, who taught with me from a draft of this book for three years, and to the student assistants who carefully reviewed the drafts to search out and correct the seemingly endless errors: Gerardo Garbulski, Andrew Kim, Julie

Ngau, Patrick Tedisch, Adam C. Powell and John Zaroulis. Thanks are also due for the helpful and thoughtful suggestions of the reviewers of the manuscript, Profs. Robert Auerback, William Bitler, Charles Brooks, and George St. Pierre. And finally, a special expression of thanks is to my wife, Kit, for her patience and constant encouragement during the preparation of this text.

David V. Ragone
Cambridge, Massachusetts
March 1994

Contents

Chapter 2   Statistical Thermodynamics   27

Chapter 3   Defects in Solids   67

Chapter 6   Transformations   165

Chapter 7   Reaction Kinetics   191

Chapter 8   Nonequilibrium Thermodynamics   221

Chapter 1

Thermodynamics: Review

This chapter reviews briefly the aspects of thermodynamics covered in Volume I of this text that are particularly important in the subjects discussed in Volume II.

The science of thermodynamics is concerned with heat and work, and transformations between the two. It is based on two laws of nature, the first and second laws of thermodynamics.

1.1   THE FIRST LAW

The first law of thermodynamics is simply the principle of conservation of energy: that energy can be neither created nor destroyed.

1.1.1   System and Surroundings

To apply the first law, the universe is divided into two parts, the system and the surroundings. The system is any portion of space or matter set aside for study. The "surroundings" is everything else. A system may be open or closed. In an open system, matter is allowed to flow into or out of the system. In a closed system, no matter enters or leaves.

1

1.1.2   Heat and Work

Energy transferred between the system and the surroundings is divided into two categories, *heat* (*Q*) and *work* (*W*). Heat is energy transferred between the system and surroundings *because of a temperature difference.* Work is all other forms of energy transferred between the system and surroundings. The algebraic sign of the heat term is positive when the heat flows from the surroundings into the system. The work term is positive when work is done on the system by the surroundings.

1.1.3   Reversibility

A process is called reversible if the initial state of the system can be restored with no observable effects in the system or the surroundings.

1.1.4   Internal Energy

The energy of a system can be divided into three categories, internal energy, kinetic energy, and potential energy. In Volume II of this text we are concerned primarily with changes in materials, independent of kinetic and potential energy terms. Thus, the kinetic and potential energy terms are omitted from energy balances.

The internal energy of a system depends on the inherent qualities, or properties of the materials in the system, such as composition and physical form, as well as environmental variables (temperature, pressure, electric field, magnetic field, etc.).

1.1.5   State Functions

The entire structure of thermodynamics is built on the concept of equilibrium states and the postulate that the change in the value of thermodynamic quantities, such as internal energy, between two equilibrium states does not depend on the thermodynamic path that the system took to get between the two states. The change is defined by the final and initial equilibrium states of that system. The internal energy is a state function, therefore:

$$\Delta U = U_2 - U_1 \qquad (1.1)$$

In contrast to internal energy, work and heat are *not* state functions. Their values in a transformation depend on the path taken.

1.1.6   The First Law

The first law of thermodynamics for a closed system that undergoes no changes in kinetic or potential energy is, in differential form:

$$\delta Q + \delta W = dU \qquad (1.2)$$

The notation δ is used with Q and W to remind us that those two quantities are not state functions.

1.1.7 Enthalpy

When a material is heated or cooled at constant pressure, and no work is done on or by it other than the pressure–volume work associated with expansion or contraction, then:

$$\delta Q = dU + P\, dV = d(U + PV) = dH \tag{1.3}$$

where H, defined as $U + PV$, is the enthalpy of the system.

1.1.8 Intensive and Extensive Properties and Notation

The properties of a system, or a portion of it, are either intensive or extensive depending on whether they are or are not functions of the size of the system. *Extensive* properties depend on the size of the system; *intensive* properties do not.

In this text, an underlined thermodynamic property signifies a *specific* property, that is, a property per unit of mass, or per mole when dealing with solutions or chemical reactions.

$$\underline{U} = \frac{U}{m} \tag{1.4}$$

The internal energy, U, is an extensive property. The specific internal energy, $\underline{U}$, is an intensive property.

1.1.9 Heat Capacities

In general, the specific heat capacity of a material under the condition that some physical parameter, I, is held constant is:

$$C_I = T \left(\frac{\partial \underline{S}}{\partial T} \right)_I \tag{1.5}$$

The notation for heat capacity is not underlined because the term is understood to be the specific heat capacity, whose units are joules per mole-kelvin [J/(mol·K)].

Using this definition, the heat capacity of a material at constant volume is:

$$C_V = \left(\frac{\partial \underline{U}}{\partial T} \right)_V \tag{1.6}$$

The heat capacity at constant pressure is:

$$C_P = \left(\frac{\partial \underline{H}}{\partial T}\right)_P \tag{1.7}$$

1.1.10   Ideal Gas

The equation of state for an ideal gas, relating the pressure (P), temperature (T), volume (V), and the number of moles (n) is:

$$PV = nRT \tag{1.8}$$

where R is the universal gas constant, 8.314 J/(mol·K).

1.1.11   Enthalpy of Formation

Enthalpy, like internal energy, is a state function. Hence, the enthalpy change between two states, ΔH, is:

$$\Delta H = H_2 - H_1 \tag{1.9}$$

In tabulations, the enthalpies of *elements* in their equilibrium states at 298 K and one atmosphere pressure are taken as zero. The enthalpies of formation of a compound ($\Delta \underline{H}_{f,298}$) at 298 K is the value of the enthalpy change when a mole of the compound is formed from its constituent elements at 298 K. The enthalpy change in a chemical reaction at 298 K is the sum of the enthalpies of formation of the products less the enthalpies of formation of the reactants at 298 K.

$$\Delta H_{298} = \sum_{\text{products}} n_p \Delta \underline{H}_{f,p,298} - \sum_{\text{reactants}} n_r \Delta \underline{H}_{f,r,298} \tag{1.10}$$

where n_p and n_r are the number of moles of the products and reactants, respectively, and the $\Delta \underline{H}_{f,298}$ terms refer to the molar enthalpies of formation of the compounds or elements involved.

The enthalpy change in a chemical reaction at a temperature T is:

$$\Delta H_T = \sum_{\text{products}} n_p \Delta \underline{H}_{f,p,T} - \sum_{\text{reactants}} n_r \Delta \underline{H}_{f,r,T} \tag{1.11}$$

The specific enthalpy of a material at temperature T relative to 298 K is:

$$\underline{H}_T = \underline{H}_{298} + \int_{298}^{T} C_P dT \tag{1.12}$$

+ the enthalpy effect of phase changes between 298 K and T

1.2 ENTROPY AND THE SECOND LAW

A change in entropy (dS) of a system as a result of heat transferred into or out of the system is defined as:

$$dS = \frac{\delta Q_{rev}}{T} \tag{1.13}$$

where δQ_{rev} is the heat transferred reversibly at temperature T. Entropy, S, is a state function. The second law for a closed system is:

$$dS = \frac{\delta Q_{actual}}{T} - \frac{\delta lw}{T} \tag{1.14}$$

where δlw, the lost work or irreversibility of the process, represents the difference between the reversible work that could have been done in a process and the actual work that was done.

1.2.1 Entropy Not Conserved

Entropy, in contrast to energy, is, in general, *not* conserved in natural processes. It is conserved only in *reversible* processes.

1.2.2. Entropy Changes

The specific entropy change for a material heated to between temperatures T_1 and T_2, at constant pressure, assuming that there are no phase changes between the two temperatures, is:

$$\underline{S}_2 - \underline{S}_1 = \int_{T_1}^{T_2} \frac{C_P}{T}\, dT \tag{1.15}$$

The specific entropy change in a phase change, such as melting ($\Delta \underline{S}_m$) may be calculated at the melting temperature when the process is *reversible*. At constant pressure, the heat added to a system to melt a mole of a material is $\Delta \underline{H}_m$, sometimes written as L (the latent heat of melting per mole) thus:

$$\Delta \underline{S}_m = \frac{\Delta \underline{H}_m}{T_m} = \frac{L}{T_m} \tag{1.16}$$

To calculate entropy changes for *irreversible* reactions, such as the freezing of a liquid below its equilibrium melting point, it is necessary to define a reversible path between the initial and final states of the system and to calculate the entropy changes along that path. If, in the case above, the material solidifies at $T < T_m$, then:

$$\Delta \underline{S}_T = \int_T^{T_m} \frac{C_{P,l}}{T} \, dT + \frac{L}{T_m} + \int_{T_m}^{T} \frac{C_{P,s}}{T} \, dT \tag{1.17}$$

This can also be written as:

$$\Delta \underline{S}_T = \frac{L}{T_m} + \int_T^{T_m} \frac{\Delta C_P}{T} \, dT$$

where $\Delta C_P = C_{P,l} - C_{P,s}$.

If the heat capacities of the liquid and solid are equal ($\Delta C_P = 0$), then the entropy of solidification will not vary with temperature. The same will be true of the enthalpy of melting, L. In that case, the Gibbs free energy of melting is:

$$\Delta \underline{G}_m = L - T\left(\frac{L}{T_m}\right) = L\left(1 - \frac{T}{T_m}\right) = \frac{L}{T_m}(T_m - T)$$

1.2.3   Entropy Changes in Chemical Reactions and the Third Law

The third law of thermodynamics states that in any chemical reaction among only *pure, crystalline* elements or stoichiometric compounds, the change of entropy is zero at the absolute zero of temperature.

$$\Delta S_0^\circ = 0 \tag{1.18}$$

The superscript degree sign indicates that we are dealing with materials in their standard states—in this case pure, crystalline substances or stoichiometric compounds at standard pressure, one atmosphere. The subscript indicates the temperature, zero kelvin.

Standard entropies at 298 K are tabulated and are equal to:

$$\underline{S}_{298} = \int_0^{298} \frac{C_P}{T} \tag{1.19}$$

+ entropy effect of phase changes between 0 and 298 K

The entropy change in a chemical reaction at 298 K is:

$$\Delta S_{298} = \sum_{products} n_p \underline{S}_{p,298} - \sum_{reactants} n_r \underline{S}_{r,298} \tag{1.20}$$

1.3 PROPERTY RELATIONS

Based on the first and second laws, the relationships among thermodynamic quantities can be determined.

$$dU = T\,dS + \delta w_{rev} \qquad \textbf{(1.21)}$$

where δw_{rev} represents all the work terms, including compression (or expansion), surface, electrical, and stress effects.

$$\delta w_{rev} = -P\,dV + \gamma\,dA + \varepsilon\,dq + F\,dl + \cdots \qquad \textbf{(1.22)}$$

If, for convenience, we consider the P-V term to represent all of the work terms, then:

$$dU = T\,dS - P\,dV \qquad \textbf{(1.23)}$$

Combining with the definition of enthalpy ($H = U + PV$), we have

$$dH = T\,dS + V\,dP \qquad \textbf{(1.24)}$$

1.3.1 Free Energies

The Helmholtz (F) and Gibbs (G) free energies are defined as follows:

$$F \equiv U - TS \qquad \textbf{(1.25)}$$

and

$$G \equiv H - TS \qquad \textbf{(1.26)}$$

Then:

$$dF = -S\,dT - P\,dV \qquad \textbf{(1.27)}$$

and

$$dG = -S\,dT + V\,dP \qquad \textbf{(1.28)}$$

1.3.2 Maxwell Relations

If dZ is an exact differential:

$$dZ = M\,dx + N\,dy$$

then

$$\left(\frac{\partial M}{\partial y}\right)_x = \left(\frac{\partial N}{\partial x}\right)_y \qquad (1.29)$$

This is true because the order of differentiating the function Z with respect to x and y does not matter.

This technique may be applied to the differentials of U, H, F, or G (Eqs. 1.23, 1.24, 1.27, 1.28). Using the expression for dG as an example:

$$-\left(\frac{\partial S}{\partial P}\right)_T = \left(\frac{\partial V}{\partial T}\right)_P \qquad (1.30)$$

1.3.3   Chemical Potentials

We can account for the addition of materials (n_i) to the system by adding terms to the equations above. For example, we can write an expression for G in terms of T, P, and n_i as follows:

$$G = G(T, P, n_i)$$

$$dG = \left(\frac{\partial G}{\partial T}\right)_{P,n} dT + \left(\frac{\partial G}{\partial P}\right)_{T,n} dP + \sum_i \left(\frac{\partial G}{\partial n_i}\right)_{T,P,n_j \neq n_i} dn_i \qquad (1.31)$$

$$dG = -S\,dT + V\,dP + \sum_i \left(\frac{\partial G}{\partial n_i}\right)_{T,P,n_j \neq n_i} dn_i \qquad (1.32)$$

The chemical potential μ_i is defined as follows:

$$\mu_i = \left(\frac{\partial G}{\partial n_i}\right)_{T,P,n_j \neq n_i} \qquad (1.33)$$

Then:

$$dG = -S\,dT + V\,dP + \sum_i \mu_i dn_i \qquad (1.34)$$

Similar expressions can be derived for the Helmholtz free energy, enthalpy, and internal energy. For example,

$$dF = -S\,dT - P\,dV + \sum_i \left(\frac{\partial F}{\partial n_i}\right)_{T,V,n_j \neq n_i} dn_i \qquad (1.35)$$

It can be demonstrated that these chemical potentials are all equal, that is:

$$\left(\frac{\partial G}{\partial n_i}\right)_{T,P,n_j \neq n_i} dn_i = \left(\frac{\partial F}{\partial n_i}\right)_{T,V,n_j \neq n_i} dn_i = \left(\frac{\partial H}{\partial n_i}\right)_{S,P,n_j \neq n_i} dn_i = \left(\frac{\partial U}{\partial n_i}\right)_{S,V,n_j \neq n_i} dn_i = \mu_i$$

$$(1.36)$$

If the component i is an ion with valence z', then the electrochemical potential, $\tilde{\mu}_i$ is:

$$\tilde{\mu}_i = \mu_i + z'\mathscr{F}\,\phi$$

where $\mathscr{F}$ is the Faraday constant and ϕ is the electric potential.

1.3.4   Partial Molar Quantities

A partial molar quantity is the partial derivative of that quantity with respect to the amount of a material, usually measured in moles, at constant temperature and pressure, and the amount of all other materials in the system. For example, the partial molar enthalpy of a material i is:

$$\overline{H}_i = \left(\frac{\partial H}{\partial n_i}\right)_{T,P,n_j \neq n_i} \tag{1.37}$$

It is important to note that the partial molar quantities for a material are not, except for the case of Gibbs free energy, equal to the chemical potential.

1.3.5   Other Definitions

The volumetric thermal expansion coefficient of a material, α_V, is defined as follows:

$$\alpha_V \equiv \frac{1}{V}\left(\frac{\partial V}{\partial T}\right)_P \tag{1.38}$$

The isothermal compressibility of a material, β_T, is:

$$\beta_T \equiv -\frac{1}{V}\left(\frac{\partial V}{\partial P}\right)_T \tag{1.39}$$

1.4   EQUILIBRIUM

Two states are in equilibrium if no reversible work can be done by having the system change between those two states. States 1 and 2 are in equilibrium if

$$\delta w_{\mathrm{rev}\,1 \to 2} = 0 \tag{1.40}$$

If there are no kinetic or potential energy differences between the two states then:

$$\delta w_{\mathrm{rev}} = G_2 - G_1 = 0 \tag{1.41}$$

For a material i existing in the two states at constant pressure (material i may be one of several components):

$$\overline{G}_{i,1} = \overline{G}_{i,2} \qquad \text{or} \qquad \mu_{i,1} = \mu_{i,2} \qquad \qquad \textbf{(1.42)}$$

In the case of a closed system at constant volume:

$$\left(\frac{\partial F}{\partial n_i}\right)_{T,V,n_j \neq n_i, \text{state1}} = \left(\frac{\partial F}{\partial n_i}\right)_{T,V,n_j \neq n_i, \text{state2}} \qquad \textbf{(1.43)}$$

or

$$\mu_{i,1} = \mu_{i,2} \qquad \qquad \textbf{(1.44)}$$

For a first-order transition in a pure material (one for which the first derivatives of G are discontinuous), the slope of the pressure–temperature line between the two phases (Clapeyron equation) is:

$$\frac{dP}{dT} = \frac{1}{\Delta \underline{V}} \left(\frac{\Delta \underline{H}}{T}\right) \qquad \qquad \textbf{(1.45)}$$

where $\Delta \underline{V}$ is the molar volume change and $\Delta \underline{H}$ is the molar enthalpy change in the transition.

In the case of evaporation of a condensed phase to an ideal gas:

$$d(\ln P) = - \left(\frac{\Delta \underline{H}}{R}\right) d\left(\frac{1}{T}\right) \qquad \qquad \textbf{(1.46)}$$

1.5   THERMODYNAMIC ACTIVITY

The *thermodynamic activity* of a material is defined as the ratio of its fugacity f to the fugacity in its standard state $f°$:

$$a_i \equiv \frac{f_i}{f°_i} \qquad \qquad \textbf{(1.47)}$$

Fugacity for a gas is defined as follows:

$$d\overline{G}_i = RT \, d \ln f_i \qquad \qquad \textbf{(1.48)}$$

where the ratio $f/p = 1$ as $p \to 0$.

The change of partial molar Gibbs free energy (or chemical potential) between two states is:

$$\overline{G}_{i,2} - \overline{G}_{i,1} = RT \ln \left(\frac{f_{i,2}}{f_{i,1}}\right) = \mu_2 - \mu_1 \qquad \qquad \textbf{(1.49)}$$

Referred to the standard state:

$$\overline{G}_i - \underline{G}_i^{\circ} = RT \ln \left(\frac{f_i}{f_i^{\circ}}\right) = RT \ln a_i \tag{1.50}$$

For a pure component, the Gibbs free energy change of *melting,* the difference in chemical potential between pure liquid and pure solid, is:

$$\Delta \underline{G}_{\text{melting}} = \frac{L}{T_m} (T_m - T) = \mu_l^p - \mu_s^p = RT \ln \left(\frac{a_l}{a_s}\right) \tag{1.51}$$

$$\mu_s^p = \mu_l^p - \frac{L}{T_m} (T_m - T) \tag{1.52}$$

where T_m = melting temperature
      L = enthalpy of melting (latent heat of fusion) per mole
μ_l^p *and* μ_s^p = chemical potentials of pure liquid and pure solid
   a_l and a_s = thermodynamic activities of pure liquid and pure solid

1.6   CHEMICAL EQUILIBRIUM

For a chemical reaction:

$$bB + cC = dD + eE$$

$$\Delta G = \Delta G^{\circ} + RT \ln J_a \tag{1.53}$$

where $J_a = \dfrac{a_D^d a_E^e}{a_B^b a_C^c}$

At equilibrium $\Delta G = 0$, hence:

$$\Delta G^{\circ} = -RT \ln(J_{a,\text{equilibrium}}) = -RT \ln K_a \tag{1.54}$$

The temperature variation of the equilibrium constant (K_a) is:

$$d(\ln K_a) = - \left(\frac{\Delta H}{R}\right) d \left(\frac{1}{T}\right) \tag{1.55}$$

1.7   ELECTROCHEMICAL CELLS

The electric potential across a cell (ε), in terms of the Gibbs free energy change of the overall reaction, is

$$\Delta G = -\varepsilon z \mathcal{F} \tag{1.56}$$

where z is the moles of electrons transferred in the overall reaction, and $\mathcal{F}$ is the Faraday constant, 96,480 coulombs per mole (C/mol).

With all the materials in their standard states:

$$\Delta G^\circ = -\varepsilon^\circ z \mathscr{F}$$ (1.57)

When materials in the cell are not in their standard states:

$$\varepsilon = \varepsilon^\circ - \left(\frac{RT}{z\mathscr{F}}\right) \ln J_a$$ (1.58)

1.8   SOLUTIONS

The thermodynamic characteristics of solutions determine their phase behavior. This property is illustrated in Sections 1.8.1–1.8.5.

1.8.1   Ideal Solutions

For an ideal solution:

$$f_i = x_i f_{i,\text{pure}}$$ (1.59)

where x_i is the mole fraction of the component i.
    If the standard state is taken to be the pure material, then:

$$f_i = x_i f_i^\circ$$ (1.60)

and the activity of i is equal to its mole fraction in the solution,

$$a_i = x_i$$ (1.61)

    The Gibbs free energy of mixing per mole in the A–B binary system (ideal solution) is:

$$\Delta \underline{G}^{\text{m}} = RT(x_{A,l}\ln x_{A,l} + x_{B,l}\ln x_{B,l})$$ (1.62)

    A graph of Eq. 1.62 is shown in Figure 1.1.
    Based on the form of Figure 1.1, it is apparent that A and B are completely miscible, unless interrupted by melting or solidification.

1.8.2   Immiscibility

In the A–B binary system, if the Gibbs free energy of mixing values for two phases (α and β) have the form shown in Figure 1.2 at a specific temperature, then the two phases are immiscible at that temperature. The limits of solubility x_α and x_β are determined by the common tangent to the α and β curves.
    The solubility of an unstable phase is greater than the solubility of a stable phase.

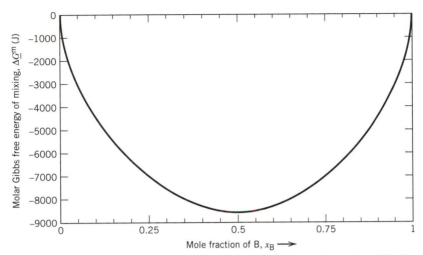

Figure 1.1 Molar Gibbs free energy of mixing of an ideal solution at $T = 1500$ K.

Figure 1.3 plots the Gibbs free energies of mixing for phases α, β, and γ as a function of composition in an A–B binary system at a specific temperature. The γ phase is unstable relative to the β phase because the Gibbs free energy of the former is greater. The common tangent between the α and γ curve intersects the α curve to the right of the α–β common tangent. Hence the solubility of the unstable γ phase is higher than that of the β phase.

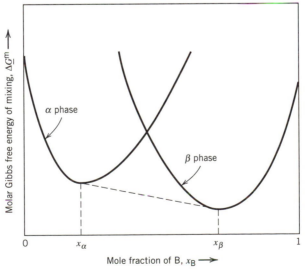

Figure 1.2 Molar Gibbs free energy of mixing for α and β phases, showing immiscibility between x_α and x_β.

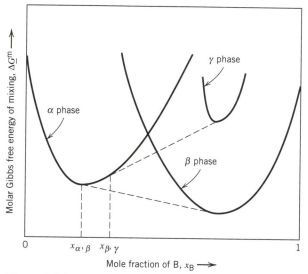

Figure 1.3 Molar Gibbs free energy of mixing for α, β, and γ phases, showing that the solubility of γ (unstable phase) is more soluble than the β phase.

1.8.3 Regular Solutions

The equation describing the Gibbs free energy of mixing per mole for a regular solution is:

$$\Delta \underline{G}^{m} = \omega x_{A} x_{B} + RT(x_{A} \ln x_{A} + x_{B} \ln x_{B}) \tag{1.63}$$

The last term in Eq. 1.63, $RT(x_{A} \ln x_{A} + x_{B} \ln x_{B})$, describes the Gibbs free energy of mixing for an ideal solution. The first term, $\omega x_{A} x_{B}$, represents that nonideality of the mixture. If the ω term is positive, then heat of mixing is positive, and the molar Gibbs free energy of the solutions of A and B can have minima as illustrated in Figure 1.4. Mixtures of A and B with the overall composition falling between $x_{B,1}$ and $x_{B,2}$ can minimize their molar Gibbs free energies by forming two solutions of composition $x_{B,1}$ and $x_{B,2}$ (i.e., becoming partially immiscible). The location of $x_{B,1}$ and $x_{B,2}$ is determined using the common tangent method described in Section 1.8.2. In the case of regular solutions, whose Gibbs free energy of mixing is symmetrical around the midpoint composition, we can locate the minima in the Gibbs free energy of mixing function by finding the compositions at which the partial derivative with respect to x_{B} is zero (recall $x_{A} = 1 - x_{B}$):

$$\left(\frac{\partial \Delta \underline{G}^{m}}{\partial x_{B}} \right)_{T} = RT \ln \left(\frac{1 - x_{B}}{x_{B}} \right) + \omega(1 - 2x_{B}) = 0 \tag{1.64}$$

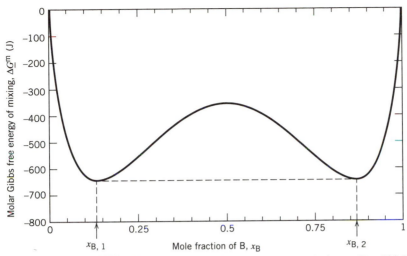

Figure 1.4 Molar Gibbs free energy of mixing for a regular solution at $T = 800$ K and $\omega = 17{,}000$ J/mol.

The solution of this equation when plotted against temperature is the phase diagram for the A–B system (Fig. 1.5). In this phase diagram, the single-phase region above T_c indicates that the two materials, A and B, are completely miscible above that temperature. At temperatures below T_c, the solution separates into two phases, α_1 and α_2. At points under the miscibility gap, the phase compositions vary with temperature. For example, at T_3 the phase compositions $x_{B,3}$ and $x_{B,4}$ are in equilibrium.

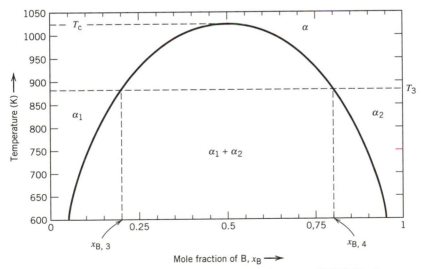

Figure 1.5 Phase diagram for a regular solution, $\omega = 17{,}000$ J/mol.

1.8.4   Spinodal Points

Another feature of phase behavior is related to the points of inflection in the curves of Gibbs free energy of mixing with respect to composition (i.e., when their second derivative is equal to zero: Eq. 1.65). These inflection points, called spinodal points, have a special significance in the study of phase transformations. The locus of spinodal points can be indicated as a dotted line in phase diagrams as in Figure 1.6.

$$\left(\frac{\partial^2 \Delta G^{\mathrm{m}}}{\partial x_B^2}\right)_T = RT\left(\frac{1}{x_A} - \frac{1}{x_B}\right) - 2\omega = 0 \qquad (1.65)$$

To appreciate the importance of the spinodal curve, consider the region to the right of the spinodal point in Figure 1.7, where the Gibbs free energy of mixing curve is concave downward. In this region the solution may begin the process of decomposition into the equilibrium phases by incremental changes in composition without increasing the total Gibbs free energy of the system. That is, fluctuations in composition can intensify without increasing Gibbs free energy. A different situation exists in the region to the left of the spinodal point (inflection point). Here, as the material attempts to separate into two phases, the Gibbs free energy of the system must increase before it can finally decrease. This difference in path for the Gibbs free energy during decomposition results in a difference in phase transformation behavior. To the left of the spinodal point the transformation is discontinuous. To the right it is not.

The top of the miscibility gap (critical mixing) is at the point where, in Eq. 1.65, $x_A = x_B$, and $T_c = \omega/2R$.

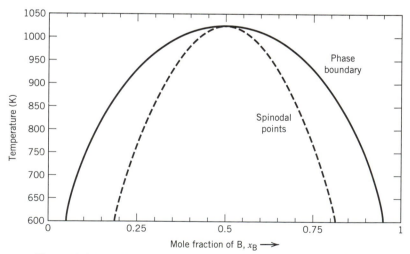

Figure 1.6 Phase boundary and spinodal points for a regular solution.

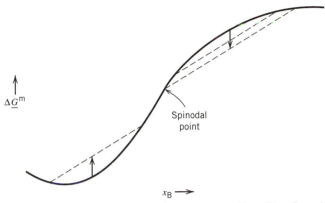

Figure 1.7 The relationship of $\Delta\underline{G}^m$ to x_B on either side of a spinodal point.

1.8.5   Gibbs–Duhem Equation

In a two-component solution, the change of Gibbs free energy of one component relative to the other as composition is changed is:

$$x_A d\overline{G}_A + x_B d\overline{G}_B = 0 \text{ (Gibbs–Duhem)}$$

or

$$x_A d\mu_A + x_B d\mu_B = 0 \tag{1.66}$$

In terms of thermodynamic activity:

$$x_A d(\ln a_A) + x_B d(\ln a_B) = 0 \tag{1.67}$$

In solutions, if solute B has a constant activity coefficient γ° (i.e., obeys Henry's law), then the solvent A behaves as if it were an ideal solution (Raoult's law).

1.9   PHASE RULE

The degrees of freedom (F) in a system are related to the number of phases present (P), and the number of components (C), as follows:

$$P + F = C + 2 \tag{1.68}$$

In the case of condensed systems where pressure is not a significant variable,

$$P + F = C + 1 \tag{1.69}$$

The number of components, C, in a system is equal to the number of chemical entities in the system, N, less the number of relations among them, R:

$$C = N - R \tag{1.70}$$

PROBLEMS

1.1 Use the data below to perform the calculations requested.

(a) Pure hydrogen at 298 K is burned with dry air at 298 K. Calculate the adiabatic flame temperature (AFT) if 200% excess oxygen (three times as much as needed stoichiometrically) is used in the torch.

(b) At the AFT in part a, how complete is the reaction of hydrogen to water vapor? That is, what fraction of the hydrogen remains unreacted?

(c) Suppose hydrogen at 298 K is burned with the stoichiometric amount of pure oxygen at 298 K. What would be the AFT if all of the hydrogen reacts?

(d) At the temperature calculated in part c, what fraction of the hydrogen remains unreacted?

(e) Taking the extent of reaction into account, what is the AFT when hydrogen at 298 K is burned with the stoichiometric amount of oxygen at 298 K?

DATA

For $H_2 + \frac{1}{2}O_2 = H_2O$
$\Delta G° = -246,000 + 54.8T$ (in joules)

Gas	Heat Capacity [J/(mol·K)]
N_2	35
O_2	35
H_2O	45
H_2	28

Note: Assume heat capacities constant.

1.2 The melting of material A may be expressed

$$A(s) = A(l)$$

Use the following notation:

$$
\begin{aligned}
T_m &= \text{equilibrium melting temperature} \\
\Delta H_m &= \text{enthalpy of fusion at } T_m \\
\Delta H_m(T) &= \text{enthalpy of fusion at } T \\
C_{p,l} &= \text{heat capacity of liquid A (assume constant)} \\
C_{p,s} &= \text{heat capacity of solid A (assume constant)}
\end{aligned}
$$

(a) Derive an equation for the enthalpy of fusion as a function of temperature.
(b) Do the same for entropy of fusion [$\Delta \underline{S}_m(T)$].
(c) Derive the expression for $\Delta \underline{G}_m(T)$, the Gibbs free energy of melting as a function of T.
(d) Express the thermodynamic activity of pure liquid A as a function of temperature, using the pure solid as the standard state.

1.3 The vapor pressure of water at 25°C is 3167.2 Pa (when water vapor alone is in equilibrium with liquid water).

(a) What is the vapor pressure of water when the liquid is subjected to 10 atm pressure?
(b) What is the thermodynamic activity of water at 10 atm pressure and 25°C if pure water at 3167.2 Pa is taken as the standard state?

Note: The density of water is 1 g/cm³.

1.4 The tensile stress on a bar of iron is increased from zero to 10^8 N/m² (about 15,000 psi) under isothermal conditions at 298 K. Assume that the iron elongates elastically.

(a) Calculate the change in entropy of the iron in joules per mole-degree kelvin.
(b) Calculate the work done per mole of iron (in joules).
(c) Calculate the change in internal energy (in joules).

DATA

For Iron:

Elastic modulus of iron = 200 GN/m² = 200 × 10⁹ N/m²
Linear thermal expansion coefficient = 11.76 × 10⁻⁶ K⁻¹
Density = 7.87 g/cm³
Molecular weight = 55.85 g/mol

1.5 A process has been proposed for depositing silica (SiO_2) at a temperature of 1800 K on a wall. In this process, a mixture of finely divided silicon (Si) powder and silica powder (both initially at 298 K) is sprayed at the wall in a stream of air (initially at 298 K). The silicon reacts with the oxygen in the air to form silica. Assume that the amount of air is regulated to provide exactly the amount of oxygen needed to convert the Si to SiO_2 and that there is no heat lost during the process (i.e., it is adiabatic).

(a) What molar ratio of Si to SiO_2 is needed in the powder mixture to achieve a temperature of 1800 K?

(b) To provide instructions to the manufacturing department, convert your answer in part a to weight percent Si in the mixture.

DATA

	Enthalpy (cal/mol)	
	$\Delta H_{f,298}$	$\underline{H}_{1800} - \underline{H}_{298}$
SiO_2	−216,417	23,637
N_2	0	11,707

1.6 This problem is concerned with thermodynamic relations in the uranium–carbon system. Assume that there are only two compounds in the system, UC and UC_2, and that they are stoichiometric compounds. Assume, also, that these compounds exist on the equilibrium diagram at the temperatures of interest in this problem.

(a) What is the activity of uranium (at 2263 K) in equilibrium with UC_2 and carbon, with solid carbon and gaseous uranium ($P = 1$ atm) as the standard states?

(b) Repeat part a (at 2263 K), using pure, liquid uranium as the standard state.

(c) What is the Gibbs free energy of formation of UC_2 from carbon and liquid uranium at 2263 K?

(d) What is the heat of vaporization of liquid uranium at 2263 K?

(e) From the data given below, what are the upper and lower limits of the Gibbs free energy of formation of UC at 2263 K? Use liquid uranium and solid carbon as the standard states.

DATA

Vapor pressure of pure, liquid uranium:

$$\log_{(10)}P \text{ (atm)} = 6.027 - \frac{24{,}040}{T} \text{ (K)}$$

The equilibrium pressure of uranium above a compact consisting of uranium dicarbide and carbon is:

T (K)	P (atm)
2063	3.8×10^{-10}
2263	1.07×10^{-8}

Note: The melting point of uranium is 1132°C.

1.7 Pure carbon monoxide (CO) is burned in dry air in a burner that uses 50% excess air.

(a) Calculate the adiabatic flame temperature (AFT) if all the CO reacts to form carbon dioxide.

(b) At the AFT calculated in part a, calculate the amount of CO unreacted if equilibrium is attained in the combustion reaction.

(c) If the combustion air is preheated to 1298 K, what will be the AFT if all the CO reacts?

(d) At the AFT calculated in part c, calculate the amount of CO unreacted if equilibrium is attained in the combustion reaction.

(e) Estimate the actual AFT if the combustion air is preheated to 1298 K and the extent of the chemical reaction is taken into account.

DATA

For $CO + \frac{1}{2}O_2 \rightarrow CO_2$, $\Delta G_T^\circ = -282{,}300 + 86.81T$ (K) joules

For the purposes of this problem, the heat capacities of the gases will be assumed to be independent of temperature and will have the following values.

Gas	Heat Capacity [J/(mol·K)]
CO	37
O_2	37
CO_2	60
N_2	37

1.8 Materials A and B form ideal solutions in both solid and liquid phases. The phase diagram for the A–B system is of the form shown. Calculate the composition of the liquidus and the solidus at 900 K (in mole percent B).

DATA

Material	Melting Point (K)	Heat of Fusion (J/mol)
A	1000	8314
B	800	6651

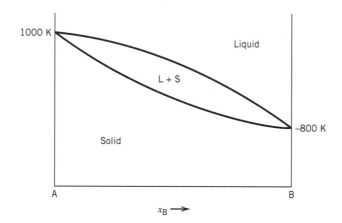

1.9 Using the accompanying phase diagram, answer the following questions for an A–B alloy having 10 mole % B.
Assuming that equilibrium is attained at all temperatures:

(a) What is the composition of the first material to solidify (in mole percent B)?

(b) What is the composition of the last liquid to solidify?

(c) What phase (or phases) is (are) present at 298 K?
During solidification, assume that diffusion in the liquid is very rapid (infinite rate) and that there is no diffusion in the solid. As the temperature drops, increasing amounts of solid are formed.

(d) Derive an equation for the composition of the liquid remaining (C_L) as a function of the fraction of the material solidified (g).

(e) What is the composition of the last liquid to solidify?

(f) What microstructures will be present at 298 K? Give amounts of each. List as primary α, primary β (if any), and eutectic.

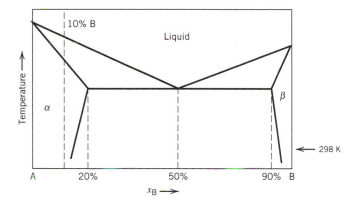

1.10 In each of the accompanying two-component phase diagrams, identify an error (in terms of the phase rule where applicable).

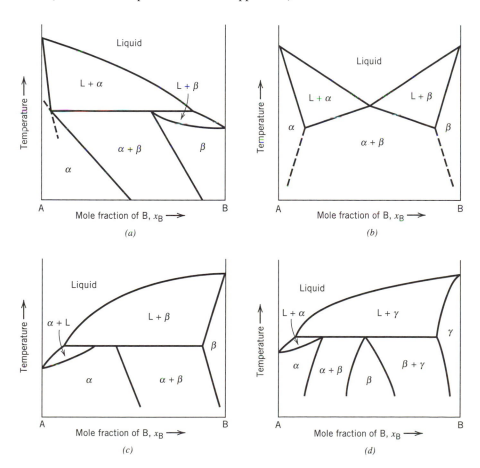

1.11 The phase diagrams for three binary systems, A–B, B–C, and C–A, exhibit simple eutectic behavior, with no solid solubility: A, B, and C form a simple ternary eutectic at a temperature of 400°C. Liquidus isotherms are shown as dashed lines in the accompanying plan (top) view of the phase diagram for the A–B–C ternary system.

The overall composition of an A–B–C solution contains 70% A, 10% B, and 20% C.

(a) Locate the overall composition on the diagram (label as P).
(b) Upon cooling from 1000°C, at what temperature will solidification begin?
$T =$ _____
(c) What is the composition of the first solid to form?
_____% A, _____% B, _____% C
(d) Trace the path followed by the liquid during solidification.
(e) What is the composition of the last material to solidify?
_____% A, _____% B, _____% C
(f) If a material that is 20% A, 15% B, and 65% C is heated from 300°C, at what temperature will melting first take place.
$T =$ _____

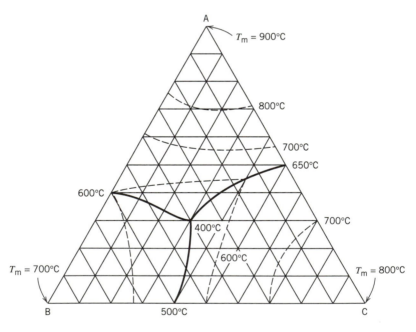

1.12 The triple point represents the temperature (and pressure) at which the three phases (ice, liquid, and vapor) exist in equilibrium. The triple-point temperature of water is 273.16 K, and the pressure is about 0.006 atm. The melting point of ice is given as 273.15 K, a value 0.001 K lower than the triple point.

The melting point of ice is the temperature at which ice and water are at equilibrium in air under *one atmosphere pressure*.

Explain in quantitative terms the difference between the triple point and the melting point.

DATA

Density of liquid water at 273.16 K = 1.000 g/cm³
Density of ice at 273.16 K = 0.917 g/cm³
Enthalpy of melting of ice = 6010 J/mol
Gas solubility in water:

Oxygen: 3.59×10^{-5} mole fraction at 273.16 K under 1 atm oxygen pressure
Nitrogen: 1.91×10^{-5} mole fraction at 273.16 K under 1 atm nitrogen pressure

Note: Assume that the gases dissolve in molecular form.

Chapter 2

Statistical Thermodynamics

This chapter consists of a brief review of the subject matter covered in Chapter 10 of Volume I, followed by a discussion of the statistical thermodynamic approach to some kinetic phenomena and to rubber elasticity.

There are two approaches to the study of thermodynamics, macroscopic and microscopic. Macroscopic thermodynamics, exemplified by the approach used in Chapters 1 to 9 of Volume I, is concerned with the relative changes among the macroscopic properties of matter, such as heats of transformation, pressure, temperature, heat capacity, density, and vapor pressure. Except in the study of chemical reactions, where it is recognized that elements combine in simple proportions to form compounds, macroscopic thermodynamics does not require any knowledge of the atomistic nature of matter. Microscopic thermodynamics, on the other hand, attempts to compute absolute values of thermodynamic quantities based on a statistical averaging of the properties of individual atoms and molecules such as the mass and volume of atoms, molecular bond strengths, and vibration frequencies. Because of this statistical averaging, microscopic thermodynamics is usually called statistical thermodynamics. Macroscopic thermodynamics can ignore the existence of molecules, molecular complexity, and quantum mechanics. Macroscopic thermodynamics cannot predict properties of a material, but it can relate them to each other. Statistical thermodynamics, which relies heavily on quantum mechanics and a knowledge of

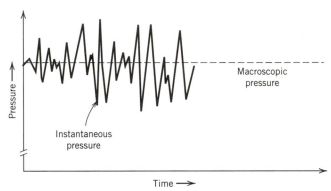

Figure 2.1 Instantaneous gas pressure (magnitude exaggerated for emphasis) as a function of time.

molecular motion and structure,[1] is used to calculate some macroscopic properties of materials from the behavior of the individual constituents of the system (atoms and molecules), and to give us a view of what is happening at the microscopic level.

This detailed picture of phenomena at the microscopic level yields additional information, beyond the calculation of the equilibrium, time-independent quantities discussed in the study of thermodynamics. It forms the basis for the study of rates of movement, such as mass diffusion, thermal conductivity, electrical conductivity, chemical reaction rates, and other kinetic phenomena.

A basic idea embodied in statistical thermodynamics is that even when a material is in equilibrium on a macroscopic scale, it is dynamic on a microscopic scale. For example, the macroscopic properties of a gas at equilibrium under constant pressure and temperature do not change with time. The state variables of macroscopic thermodynamics (T, V, U, etc.) are fixed. On the microscopic scale, however, the atoms or molecules are in motion, and their configuration changes constantly. As an example, consider the pressure exerted by a gas in a container. The observed pressure of gas results from the force the molecules exert on the container wall when the particles strike and bounce back from it. If we were able, by some means, to measure the instantaneous gas pressure in extremely short time intervals, we would observe a rapidly fluctuating pressure. What in macroscopic thermodynamics is called "pressure" is the average of this fluctuating pressure over time. This is illustrated schematically in Figure 2.1, in which the pressure fluctuations have been exaggerated for emphasis.

We need not have a detailed picture of the motion of each molecule in a system

---

[1]The treatment of statistical thermodynamics used in this chapter relies on the ideas of quantum mechanics, such as quantized energy levels. But, the idea of quantization is not essential to statistical thermodynamics. The ideas that form the basis of classical (non–quantum mechanical) statistical thermodynamics predate the advent of quantum mechanics.

to gain some interesting information about the macroscopic properties of gases or solids. A statistical approach will suffice. To use this approach, some basic notions are needed.

2.1   MACROSTATES AND MICROSTATES

To establish the basis for the study of statistical thermodynamics, one based on probabilities, we will distinguish between a macrostate and a microstate. A *macrostate* of a system is what, in classical thermodynamics, is called a "state" and is characterized by a few state variables, such as temperature, volume, and internal energy. A *microstate* of a system characterizes the state of all the particles in the system. For example, a gas in an isolated system at equilibrium at a given volume is in one macrostate. To specify a microstate of this gas, we must specify the position and velocity of all the molecules in the system within the limits imposed by the uncertainty principle.

A system in one *macrostate* passes, with time, through many *microstates*. In fact it passes very rapidly through many microstates, because atoms and molecules move and change direction rapidly. An atom in a solid, for example, vibrates at a frequency on the order of 10^{13} times per second. Gas molecules have velocities of several hundreds of meters per second. The picture of a material as a rapidly changing system leads us to the realization that when we observe a property of a system, we are really seeing the average of this property in all the microstates the material passed through during the observation time. This is one of the basic ideas in statistical thermodynamics: *the property of a macrostate is an average over the properties of all the microstates a system goes through.*

An important premise in statistical thermodynamics is that a system in a given macrostate can exist in every microstate consistent with the constraints of the macrostate. The *ergodic hypothesis* of microscopic thermodynamics states that *the time average of the properties of a system is equivalent to the instantaneous average over the ensemble of the microstates available to the system.*

To compute the macroscopic average of a property we need to know:

- the property of each microstate
- which microstates the system can be in
- the probability that the system will be in a given microstate

To illustrate some of these concepts, consider a system consisting of three particles, each of which may have one of three energies: I, II, or III, corresponding to 0, 0.1, or 0.2 electron volt. Table 2.1 lists all the possible configurations of the system. The quantity Ω represents the number of different ways a particular configuration can be achieved. For example, state B, which consists of two particles at level III and one at level II, can be achieved in three different ways. That is, level II can be occupied by any of the three particles, the other two being at level III. The

Table 2.1 Possible Distributions of Three Particles Among Three Different Energy Levels

Energy Level	Configurations									
	A	B	C	D	E	F	G	H	I	J
III at 0.2 eV	3	2	2	1	1	0	1	0	0	0
II at 0.1 eV	0	1	0	2	1	3	0	2	1	0
I at 0 eV	0	0	1	0	1	0	2	1	2	3
Total energy of configuration, eV	0.6	0.5	0.4	0.4	0.3	0.3	0.2	0.2	0.1	0
Ω (sum = 27)	1	3	3	3	6	1	3	3	3	1
			6		7		6			
P_i at 1000 K (sum = 1.00)[a]	3×10^{-4}	0.003	0.021		0.077		0.209		0.334	0.356

[a]The symbol Ω denotes the thermodynamic probability of the configuration.

value of Ω, sometimes called the thermodynamic probability,[2] is given for each configuration in Table 2.1.

For the situation just described, the expression for Ω in the case involving three energy levels is:

$$\Omega = \frac{N!}{N_I! N_{II}! N_{III}!} \qquad (2.1)$$

where N_I = the number of particles at level I

N_{II} = the number of particles at level II

N_{III} = the number of particles at level III

$N = N_I + N_{II} + N_{III}$

In general, for i levels or categories, the term Ω is:

$$\Omega = \frac{N!}{\Pi_i N_i!} \qquad (2.2)$$

where $\Pi_i N_i!$ is the product of all the $N_i!$ terms.

---

[2]This is inconsistent with the usual definition of probability, in which the sum of the probabilities is one. The two can be reconciled by dividing the microstates per macrostate by the total number of possible microstates. In the example shown in Table 2.1, each value of Ω would be divided by 27, the total number of possible microstates.

Suppose we isolate the system described in Table 2.1 so that the total energy of the system is 0.3 eV. In an isolated system with N particles in a volume V, with a fixed total energy, each configuration of the system, each microstate, must have the same energy, E. In the system described in Table 2.1, there are seven ways this can be achieved, that is, seven microstates (the sum of configurations E and F) each of which has a total energy of 0.3 eV. One of the basic assumptions in statistical thermodynamics is that *all microstates that are in the same total energy level are equally probable.* The accepted term in statistical thermodynamics for this collection of microstates with constant energy, volume, and number of particles, is *microcanonical ensemble.*

2.2   THE BOLTZMANN HYPOTHESIS

The Boltzmann hypothesis states that the entropy of a system is linearly related to the logarithm of Ω (defined in Eq. 2.2).

$$S = k \ln \Omega \qquad\qquad (2.3)$$

The general form of the Boltzmann equation can be rationalized by considering two independent systems and comparing the total entropy of the two with the total probability function for the two. The entropy of the two taken together, S, is simply the *sum* of the two entropies, $S_1 + S_2$, because entropy is an extensive property in the macroscopic sense. The situation for probability is different. For each microstate in the first, all the microstates in the second must be counted. Thus total number of microstates in the two taken together is the *product* of the number in each, $\Omega = \Omega_1 \times \Omega_2$. Boltzmann hypothesized that entropy is proportional to a function of the thermodynamic probability (macrostates per microstate). Because of the additive nature of entropy and the multiplicative nature of thermodynamic probability, the function that relates the two had to be logarithmic. That is a rationalization of the Boltzmann equation (Eq. 2.3). The Boltzmann hypothesis is best viewed, however, simply as a brilliant insight that has stood the test of many experimental verifications.

2.3   ENTROPY OF MIXING

The principle discussed in the preceding section can be used to calculate the entropy of mixing of two components of an ideal solid solution. In an ideal solution, the positioning of atoms on a lattice is random because there is no enthalpy or energy of mixing. Hence, all the configurations have the same energy, and are all equally probable.

For one mole of atoms (the sum of N_1 and N_2 is Avogadro's number of atoms, N_A) the function Ω is:

$$\Omega = \frac{N_A!}{N_1! N_2!} \qquad\qquad (2.4)$$

The entropy of mixing (per mole in this case), $\underline{S}_M$, sometimes called the config-urational entropy, is:

$$\underline{S}_M = k \ln \Omega = k \ln \frac{N_A!}{N_1!N_2!}$$

After applying the Stirling approximation (see Vol. I, Section A.17) and some algebraic manipulation, the expression for the entropy of mixing becomes

$$\underline{S}_M = -R(x_1 \ln x_1 + x_2 \ln x_2) \tag{2.5}$$

where x_1 and x_2 refer to the mole fractions of 1 and 2, respectively.

2.4   SYSTEMS AT CONSTANT TEMPERATURE AND THE CANONICAL ENSEMBLE

Section 2.3 discussed the distribution of microstates and the properties of an isolated system, that is, one at constant energy. Consider now a closed system, with a fixed number of particles N and in a fixed volume V immersed in a large, constant temperature bath in which energy transfer (heat flow) to or from the system is possible. The temperature of the system will be held constant, but the energy may fluctuate around some average value, just as the pressure of a system fluctuates around its average value. We can now think of an ensemble of systems, each at temperature T, but having an instantaneous energy that is allowed to vary. This is called a *canonical ensemble*. Because the energy of the system may vary, each of the configurations in Table 2.1 is possible, but all are not equally probable. To calculate these probabilities, we need expressions for the entropy and energy of the various configurations.

An expression for the entropy of this canonical ensemble is given by

$$S = -k \sum_i P_i \ln P_i \tag{2.6}$$

where P_i is the probability that the system will be in microstate i, and the sum is the sum over all the microstates.

The logic behind this statement for entropy, sometimes called the Gibbs formulation, is discussed in many texts on statistical thermodynamics (see, e.g., Refs. 1–4). The Gibbs formulation can be shown to be identical to the Boltzmann formulation (Eq. 2.3). Either of the two statements, Gibbs or Boltzmann as appropriate, can be used to calculate equilibrium distributions and conditions by adopting the principle:

The state of knowledge we must assume is the one that maximizes S relative to the information given.

2.5   BOLTZMANN DISTRIBUTION

Consider, first, a situation in which N particles are distributed among different energy levels. Assume that there is *no limit on the number of particles that may exist at any energy level.* Note that this condition would not be met when the Pauli exclusion principle applies—that is, when the number of particles (typically electrons) in each state is limited. These cases are considered in Section 2.11 in our discussion of the Fermi–Dirac distribution.

The task is to find the distribution of N particles among the i energy levels that will maximize the entropy S, subject to the constraints that the energy E be equal to $\Sigma N_i E_i$ and $N = \Sigma N_i$. The probability P_i, which is the number of particles in the state i divided by the total number of particles (N_i/N), is the same as the probability that any specific particle will be found in state i.

The result is that the probability P_i that a particular particle will be found in its ith energy level is given by

$$P_i = \frac{\exp(-E_i/kT)}{Z} \tag{2.7}$$

where $Z = \sum_i \exp(-E_i/kT)$ and is called the partition function.

2.6 PARTITION FUNCTION

As defined here and in Section 2.5, the partition function Z is the sum over all the energy states allowed (or the sum over all energy levels accounting for degeneracy, as discussed in Section 2.7). It is a particularly useful function because all the thermodynamic properties of a system may be calculated once the partition function is known. In a system consisting of one particle, the entropy is:

$$S = k \ln Z + kT \left(\frac{\partial \ln Z}{\partial T}\right)_V \tag{2.8}$$

The Helmholtz free energy is

$$F = -kT \ln Z \tag{2.9}$$

2.7 DEGENERACY

At any energy level E_i, there may be a number of states g_i, all having the same energy, but having other characteristics that distinguish them from one another. This factor, g_i, is called the *degeneracy* or the *statistical weight* of the energy level. It is important to be clear on whether we are asking about the probability that a particle will be at a specified energy *level*, or whether it will be in a specific *state* within that level. The probability that it will be in a specific *state* is:

$$P_i = \frac{\exp(-E_i/kT)}{Z} \tag{2.7}$$

where $Z = \sum_i \exp(-E_i/kT)$, summed over all the states of the system.

The probability that a particle will be in a specified energy *level* is:

$$P_j = \frac{g_j \exp(-E_j/kT)}{Z} \qquad (2.10)$$

where $Z = \sum_j g_j \exp(-E_j/kT)$, summed over all the levels.

Equations 2.7 and 2.10 may be used interchangeably. One can sum over each possible state, considering all the substates within one energy level to be independent, in which case Eq. 2.7 applies. Alternatively, if one accounts for the number of substates in each energy level, then Eq. 2.10 applies, and we sum over energy levels. Referring to the system described in Table 2.1, the partition function evaluated as the sum over all the states of the system would have 27 terms in it, one for each microstate. The partition function evaluated as the sum over levels would have only seven terms, one for each energy level. At each energy level, the degeneracy g_j, would be equal to the number of microstates at that energy level. The B level would have a degeneracy of 3, for example.

As a numerical example, let us calculate the population of the total energy levels for the configurations in Table 2.1 (A–J) at a temperature of 1000 K. The partition function evaluated in terms of degeneracy is:

$$Z = \sum_A^J g_j \exp\left(-\frac{E_j}{kT}\right)$$

$$Z = 1 \exp\left(\frac{0}{kT}\right) + 3 \exp\left(\frac{-0.1}{kT}\right) + 6 \exp\left(\frac{-0.2}{kT}\right)$$

$$+ 7 \exp\left(\frac{-0.3}{kT}\right) + 6 \exp\left(\frac{-0.4}{kT}\right) + 3 \exp\left(\frac{-0.5}{kT}\right)$$

$$+ 1 \exp\left(\frac{-0.6}{kT}\right)$$

At a temperature of 1000 K, $Z = 2.812$.

The probability of finding the system in state C–D, with a total energy of 0.4 eV, is:

$$P_{0.4} = \frac{6 \exp(-0.4/1000 \text{ K})}{Z} = \frac{0.058}{2.812} = 0.021$$

The average energy of the system at 1000 K is:

$$\overline{E} = \sum_A^J P_j E_j = 0.108$$

2.8   DISTINGUISHABILITY OF PARTICLES

When one is considering many particles, the question of distinguishability among the particles arises. In a solid material in which atoms are localized (i.e., vibrate about some fixed location), the particles (the atoms) are distinguishable. If the atoms are noninteracting—that is, if we do not have to account for the energies of combinations of atoms—then the grand partition function for all the atoms, Φ, is the product of the partition functions of each of the individual atoms (Eq. 2.7). To understand this relationship, consider the grand partition function for two atoms. The grand partition function must take into account all combinations of the energy states of both atoms. For each state of the first, we must account for Z states of the second. Thus for the two, we have Z times Z, or Z^2 states. Taking N particles into account yields:

$$\Phi = Z^N \tag{2.11a}$$

In the case of atoms or molecules in a gas, the particles are not localized and are indistinguishable. The grand partition function for this collection of N identical, noninteracting particles is the product of the partition function of each of the particles divided by the term $N!$ to take account of the number of ways that the particles may be rearranged indistinguishably.

$$\Phi = \frac{Z^N}{N!} \quad \text{(for gases)} \tag{2.11b}$$

Table 2.2 gives some thermodynamic functions in terms of the partition function for localized and nonlocalized particles.

Table 2.2 Partition Function (Φ) for Many (N) Particles Based on Partition Function for One Particle (Z)

Indistinguishable Particles (gas)	Distinguishable Particles (solid)
$\Phi = \dfrac{Z^N}{N!}$	$\Phi = Z^N$
$\ln \Phi = N \ln \dfrac{Z}{N} + N$	$\ln \Phi = N \ln Z$
$\left(\dfrac{\partial \ln \Phi}{\partial T}\right)_v = N \left(\dfrac{\partial(\ln Z)}{\partial T}\right)_v$	$\left(\dfrac{\partial \ln \Phi}{\partial T}\right)_v = N \left(\dfrac{\partial(\ln Z)}{\partial T}\right)_v$
$S = kN \left[\ln\left(\dfrac{Z}{N}\right) + 1\right] + NkT \left(\dfrac{\partial(\ln Z)}{\partial T}\right)_v$	$S = kN \ln Z + NkT \left(\dfrac{\partial(\ln Z)}{\partial T}\right)_v$
$F = -NkT \left[\ln\left(\dfrac{Z}{N}\right) + 1\right]$	$F = -NkT \ln Z$
$P = -\left(\dfrac{\partial A}{\partial V}\right)_T = NkT \left(\dfrac{\partial(\ln Z)}{\partial V}\right)_T$	$P = -\left(\dfrac{\partial A}{\partial V}\right)_T = NkT \left(\dfrac{\partial(\ln Z)}{\partial T}\right)_v$

2.9   IDEAL GAS

To establish the partition function of an ideal monatomic gas, one in which only translation of the atom (not rotation, vibration, nor electronic excitation) need be considered, we may use the particle-in-a-box approach from quantum mechanics. The gas is considered to be contained in a cubical box of side L, and volume L^3. If we consider the velocity of the particle to be quantized, its wavelength in the x direction is defined by the DeBroglie relationship:

$$mv = \frac{h}{\lambda} \tag{2.12}$$

where $\lambda = 2L/i$ and $i = 1, 2, 3, 4, \ldots$.

The equations for velocity in the x direction and the various energy levels are

$$v_x = \frac{hi}{2mL} \tag{2.13}$$

where $i = 1, 2, 3, 4, \ldots$, and

$$E_x = \frac{h^2 i^2}{8mL^2} \tag{2.14}$$

where $i = 1, 2, 3, 4, \ldots$.

The partition function in the x direction is

$$Z_x = \left(\frac{2\pi mkT}{h^2}\right)^{1/2} L \tag{2.15}$$

The partition function in all three directions, x, y, and z, is the product of the partition function of individual partition function in each direction:

$$Z_{xyz} = Z_x Z_y Z_z = \left(\frac{2\pi mkT}{h^2}\right)^{3/2} V = \frac{V}{h^3} (2\pi mkT)^{3/2} \tag{2.16}$$

2.10   MAXWELL–BOLTZMANN DISTRIBUTION: IDEAL GAS

The energy of a monatomic gas molecule imparted by its motion (translation only) is:

$$E_{i,j,k} = \frac{h^2}{8mL^2} (i^2 + j^2 + k^2) \tag{2.17}$$

The density of states, $g(E_i)$ is (see Vol. I, Section 10.11):

$$g(E_i) = 4\sqrt{2}\ \pi\ \frac{V}{h^3}\ m^{3/2}E_i^{1/2} \tag{2.18}$$

where $g(E_i)$ is the number of states between E and $E + dE$.
The probability of occupancy of the state E_i, $P(E_i)$, is:

$$P(E_i) = g(E_i)\ \frac{\exp(-E_i/kT)}{Z} = \frac{N(E_i)}{N} \tag{2.19}$$

where $N(E_i)$ is the number of atoms in state E_i.
And the partition function Z of an ideal gas is:

$$Z = \frac{V}{h^3}\ (2\pi mkT)^{3/2} \tag{2.16}$$

Hence, the number of atoms in each energy state $[N(E_i)]$ in a mole (N) of gas is the product of the density of states and the probability of occupancy:

$$N(E_i) = 2\pi N \left(\frac{1}{\pi kT}\right)^{3/2} E_i^{1/2}\ \exp\left(-\frac{E_i}{kT}\right) dE \tag{2.20a}$$

Equation 2.20a is Maxwell–Boltzmann distribution for an ideal gas in terms of energy. To convert this to a distribution of velocities, we note that the energy of a particle is $\frac{1}{2}mv^2$. Thus:

$$N(v) = 4\pi N \left(\frac{m}{2\pi kT}\right)^{3/2} v^2\ \exp\left(-\frac{mv^2}{2kT}\right) dv \tag{2.20b}$$

The average velocity ($\langle v \rangle$) of a particle is:

$$\langle v \rangle = \frac{\displaystyle\int_0^\infty vN(v)}{N} = \left(\frac{8kT}{\pi m}\right)^{1/2} \tag{2.21}$$

2.11 FERMI–DIRAC DISTRIBUTION

The Maxwell–Boltzmann distribution (Section 2.10) assumes that there is no limit to the number of particles that may occupy a specific state defined by the values of i, j, and k; that is, these particles are not subject to the Pauli exclusion principle. In the case of "free" (conduction) electrons in a metal, this assumption is not justified. The Pauli exclusion principle, as it applies to an electron gas, states that no two electrons may have the same set of quantum numbers. Thus, the approach to statistics must be modified.

To evaluate the function, Ω in Eq. 2.3, an expression for the number of microstates per macrostate is needed. In this case, the macrostate is defined by g_i, the number of states at energy level E_i, and the number of them, N_i, that are occupied. A microstate consists of one way of arranging the N_i particles among the g_i places available. The total number of possible arrangements is equivalent in statistical terms to the number of combinations of N objects taken M at a time, which, from probability considerations, is:

$$C = \frac{N!}{(N - M)!M!}$$

Thus for the ith level with g_i states, N_i of which are occupied, we have

$$C = \frac{g_i!}{N_i!(g_i - N_i)!} \qquad (2.22)$$

The function Ω is the product of these terms for each energy level:

$$\Omega = \Pi_i \frac{g_i!}{N_i!(g_i - N_i)!} \qquad (2.23)$$

The entropy $k \ln \Omega$ is

$$S = k \sum_i [g_i\ln g_i - N_i\ln N_i - (g_i - N_i)\ln(g_i - N_i)] \qquad (2.24)$$

To derive an expression for the population of the various states N_i, S (or $\ln \Omega$) is maximized subject to two constraints:

$$\sum_i E_iN_i = U$$

$$\sum_i N_i = N$$

The result is the Fermi–Dirac distribution:

$$N_i = \frac{g_i}{1 + \exp\left(\dfrac{E_i - \mu}{kT}\right)} \qquad (2.25)$$

In many texts on solid state physics, the μ term in Eq. 2.25 is called the Fermi energy, usually denoted as E_F.

It can be shown that the Fermi–Dirac distribution (Eq. 2.25) becomes the Boltzmann distribution (Eq. 2.7) when the exponential term, $\exp[(E_i - \mu)/kT]$, is much greater than one.

2.12   EFFUSION: LANGMUIR EQUATION

Based on the concepts derived and explained earlier in this chapter, we are now in a position to answer some questions concerning kinetics, that is, questions that have to do with rates of change. This is the bonus provided by statistical thermodynamics. One such question involves the rate at which particles (atoms) strike a unit surface of a container per unit time, given the pressure and temperature of the gas in the container. The answer to this question is interesting from several points of view. It can yield an estimate of the time needed for a totally clean surface to be covered with a monolayer of atoms or molecules, assuming that all the molecules that hit the surface stick to it (become adsorbed). It also can allow us to calculate how many atoms will escape from a small hole in a vessel per unit time, given the area of the hole. This will turn out to be the basis of a method to measure the vapor pressure of materials of very low volatility. And, as we will see later in the chapter, it also provides an answer to the question of how many particles may evaporate from a surface per unit time.

To approach this problem, imagine a cylinder of length L, with cross-sectional area A, oriented with its axis along the x axis of a set of coordinates (Figure 2.2). This can be considered to be a collision cylinder; that is, it will contain atoms that will collide with a wall at the end of the cylinder within the time interval τ. The length L is equivalent to $v_x\tau$. The quantity v_x describes the velocity of a particular particle. All the atoms in the cylinder having the velocity v_x in the positive direction will strike that wall within the time interval τ. If we designate the atoms striking the wall in time τ as N^*, then:

$$N^* = \sum_i v_{x,i}\tau A \, \frac{N}{V} \frac{1}{2} \, (\text{probability of } v_{x,i})$$

The term "probability of $v_{x,i}$" is evaluated from Eq. 2.17 by setting the j and k terms equal to zero because we are interested only in the velocity in the x direction.

$$N^* = \sum_i v_{x,i}\tau A \, \frac{N}{V} \frac{1}{2} \, \frac{\exp\left(-\dfrac{h^2}{8mV^{2/3}kT} \, i^2\right)}{Z} \tag{2.26a}$$

where $v_{x,i}$ is the ith state of the velocity in the x direction, and N/V is the number of atoms in volume V.

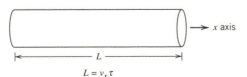

$$L = v_x\tau$$

Figure 2.2 Collision cylinder for molecules and a wall.

Substituting an integral for the sum and defining γ (a lumped parameter) as follows:

$$\gamma \equiv \frac{h^2}{8mV^{2/3}kT}$$

we have

$$\frac{N^*}{A\tau} = \frac{1}{2}\frac{N}{VZ}\gamma\left(\frac{2kT}{m}\right)^{1/2}\int_0^\infty i\,\exp(-\gamma^2 i^2)\,di \qquad \textbf{(2.26b)}$$

The integral in the preceding expression can be evaluated as a definite integral:

$$\int_0^\infty i\,\exp(-\gamma^2 i^2)\,di = \frac{1}{2\gamma^2}$$

Combining with the equation for the partition function, Z, of an ideal gas in one dimension (Eq. 2.15):

$$\frac{N^*}{A\tau} = \frac{NkT}{V}\left(\frac{1}{2\pi mkT}\right)^{1/2}$$

$$\frac{N^*}{A\tau} = \frac{P}{(2\pi mkT)^{1/2}} \qquad \textbf{(2.27)}$$

We now have an expression for the number of particles striking a wall per unit time per unit area as a function of the pressure, temperature, and the mass of the particle. This is called the Langmuir equation. Using this equation (Eq. 2.27), we can estimate the level of vacuum required to keep a surface clean for a specified period of time. To put some values in Eq. 2.27:

$$\frac{N^*}{A\tau} = \frac{P\,(\text{atm})\times 1.013\times 10^5}{\left[2\pi\dfrac{M\times 10^{-3}}{6.022\times 10^{23}}\times 1.38\times 10^{-23}T\right]^{1/2}}\times 10^{-4} \qquad \textbf{(2.28)}$$

where $N^*/A\tau$ = number of particles striking the surface per square centimeter per second

M = molecular weight (g/mol)

For the purposes of this estimation, let us assume that a monolayer of atoms on a surface will contain about 10^{15} atoms per square centimeter. For the case of oxygen (molecular weight 32 g/mol) at 300 K:

$$\frac{N^*}{A\tau} = 2.72\times 10^{23}P = \frac{10^{15}}{\tau} \qquad \textbf{(2.29)}$$

The accommodation coefficient, α, for the deposition atoms or molecules, is the ratio of the number of atoms that stick to the surface to the number that strike it. It is zero if none stick; it is one (unity) if all that hit stick. If the accommodation coefficient is one, then at a pressure of 10^{-10} atm, a monolayer will be formed in about 37 seconds. If we require the surface to be kept clean for one hour, the pressure required would be about 10^{-12} atm.

The Langmuir equation can also be used to calculate the maximum rate of evaporation of a surface. Consider a material at a temperature, T, in equilibrium with its vapor. The number of atoms of the vapor striking the surface per unit time and sticking to it is given by:

$$\frac{N^*}{A\tau} = \alpha \frac{P}{(2\pi mkT)^{1/2}} \tag{2.30}$$

If the surface is indeed at equilibrium, it does not change with time. The number of atoms that evaporate must be equal to the number of atoms that strike it and stick to it. Thus the rate of evaporation must be equal to the rate of deposition calculated by Eq. 2.30. Suppose the vapor in equilibrium with the surface were completely removed—that is, the chamber were evacuated completely. The evaporation rate from the surface would still be the value calculated by Eq. 2.30. The *maximum* rate of evaporation of the surface would occur in the case of $\alpha = 1$. Therefore the maximum rate of evaporation of a surface into a vacuum, given in terms of atoms or molecules leaving per unit area per unit time, is:

$$\frac{N}{A\tau} = \frac{P^\circ}{(2\pi mkT)^{1/2}} \tag{2.31}$$

where P° is the equilibrium pressure of the evaporating material at the surface temperature.

The relationship in Eq. 2.30 can also be used to calculate the rate at which particles (atoms or molecules) can escape through a hole in a container, given the temperature and pressure in the container, and the size of the hole. Actually, the relationship is used more frequently to calculate the pressure of the gas in the system knowing the temperature, the cross-sectional area of the hole, and the rate of mass loss. As an example, let us calculate the vapor pressure of liquid aluminum from the following information. When liquid aluminum is held in a stable, inert container suspended in a vacuum furnace ($P = 0$), the mass loss through a hole with cross-sectional area 2×10^{-3} cm^2 is 1.7×10^{-9} g/s at a temperature of 1250 K. The molecular weight of aluminum is 27 g/mol, and the vapor is monatomic.

Rearranging Eq. 2.31, and recognizing that the value of the accommodation coefficient must be one in this case, because all the atoms that strike the container wall at the hole will go through, we have

$$P° = \frac{N*}{A\tau} (2\pi mkT)^{1/2}$$

$$\frac{N*}{A\tau} = \frac{1.7 \times 10^{-9}}{27} \times 6.022 \times 10^{23} \times \frac{1}{2 \times 10^{-7}} \text{ m}^{-2} \text{ s}^{-1}$$

$$P° = \frac{N*}{A\tau} \left(2\pi \frac{27 \times 10^{-3}}{6.022 \times 10^{23}} \times 1.38 \times 10^{-23} \times 1250 \right)^{1/2}$$

$$P° = 1.32 \times 10^{-2} \text{ N/m}^2$$

$$P° = 1.30 \times 10^{-7} \text{ atm}$$

This technique, known as the Knudsen effusion method, is a convenient way to determine vapor pressures in low pressure regions, down to about 10^{-5} torr. The method, of course, requires a knowledge of the molecular weight of the effusing species.[3]

The effusion of gases from one chamber to another results in an interesting situation when the two chambers containing the same gas are held at different temperatures. Two such chambers are shown in Figure 2.3. The wall separating them has a small hole of diameter a. The chamber on the left is maintained at temperature T_1. The chamber on the right is maintained at temperature T_2. The number of gas molecules moving from the left chamber to the right per unit time ($N*/\tau$) is

$$\frac{\vec{N}}{\tau} = A \frac{P_1}{(2\pi mkT_1)^{1/2}} = \left(\frac{\pi a^2}{4} \right) \frac{P_1}{(2\pi mkT_1)^{1/2}} \tag{2.32}$$

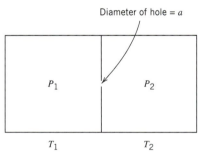

Diameter of hole = a

P_1 P_2

T_1 T_2

Figure 2.3 Pressures of a gas in connecting chambers held at different temperatures.

---

[3]Several elements, such as bismuth, phosphorus, antimony, and arsenic, have multiple gaseous species. The pressures of these elements can be measured using a torsion–effusion technique in which the effusion cell has two holes in it on opposite faces, displaced by a measured distance. The cell is mounted so that the direction of the effusion is in the horizontal plane. The offset effusions produce a torque on the cell that is proportional to the pressure of the effusing gas, independent of its molecular weight.

The number moving from the right chamber to the left is:

$$\frac{\overleftarrow{N}}{\tau} = \left(\frac{\pi a^2}{4}\right) \frac{P_2}{(2\pi mkT_2)^{1/2}} \tag{2.33}$$

Let the chambers be maintained at T_1 and T_2 until a steady state is attained, that is, until net flow ceases. At steady state:

$$\frac{\overrightarrow{N^*}}{\tau} = \frac{\overleftarrow{N^*}}{\tau}$$

$$\frac{P_1}{T_1^{1/2}} = \frac{P_2}{T_2^{1/2}} \quad \text{or} \quad \frac{P_1}{P_2} = \left(\frac{T_1}{T_2}\right)^{1/2} \tag{2.34}$$

Based on Eq. 2.34, the pressures in the two chambers, one at T_1 and the other at T_2, will be different. This is certainly not intuitively obvious. In fact, it is counter to our common experience. If we open a door between a warm house at T_2 and the colder outdoors at T_1, the indoor pressure will not differ from the outdoor pressure. The explanation lies in the difference in the types of flow that take place. The flow of gas described by the Langmuir equation is different from the flow through a doorway at atmospheric pressure. To make this distinction, we need to know something about the distance between collisions of gas molecules, which is the subject of the next section.

2.13 MEAN FREE PATH

The collision volume approach used in Section 2.12 may also be used to estimate the mean free path of a molecule in a gas, that is, the mean distance between collisions. For the purposes of this estimation, assume that all the particles (molecules) involved are spherical and of the same material. To derive an expression for the mean free path, we focus our attention on the relative motion among the particles. Imagine just one particle to be moving. This particle sweeps out a collision cylinder with a radius σ, where σ is the diameter of the atom (Figure 2.4).

All the atoms (or molecules) with centers in the collision cylinder will collide with the "projectile" atom. The cross-sectional area of the collision cylinder is

$$A = \pi\sigma^2$$

In the time interval t, the volume of the collision cylinder is:

$$V = \bar{v}t\pi\sigma^2 \tag{2.35}$$

where $\bar{v}$ is the average absolute velocity of the atom.

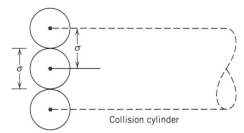

Figure 2.4 Collision cylinder for gaseous molecules.

The number of molecules in this volume is the number of collisions that will be sustained. The average distance between collisions is defined as λ, the total distance a molecule travels in a unit time divided by collisions it sustained in that time. Thus:

$$\lambda = \frac{\bar{v}t}{(N/V)\bar{v}\tau\pi\sigma^2} = \frac{1}{\pi\sigma^2(N/V)} \tag{2.36}$$

where N/V is the number of molecules per unit volume.

Actually, Eq. 2.36 does not reflect consideration of the various geometrical factors to account for the ability of atoms to approach each other at different angles. If one takes this into account, the expression becomes

$$\lambda = \frac{1}{\sqrt{2}\,\pi\sigma^2(N/V)} \tag{2.37}$$

In terms of pressures and temperatures, this becomes, for an ideal gas:

$$\lambda = \frac{kT}{\sqrt{2}\,\pi\sigma^2 P} \tag{2.38}$$

As an example, consider the motion of oxygen at 300 K. If we take the "diameter" of the oxygen molecule to be 3×10^{-10} m (3 Å), the expression in Eq. 2.38 reduces to

$$\lambda = \frac{(1.38 \times 10^{-23})(300)}{\sqrt{2}\,\pi(3 \times 10^{-10})^2(P_{(atm)} \times 1.013 \times 10^5)}$$

$$\lambda = \frac{1.02 \times 10^{-7}}{P_{atm}} \text{ meter}$$

The mean free path is inversely proportional to the pressure. At a pressure of one atmosphere, the mean free path is about 10^{-7} meter or 10^{-5} cm. At a pressure of 10^{-9} atm, the mean free path is about 100 meters (i.e., 10^4 cm). This tells us that molecules in high vacuum chambers of normal dimensions do not often strike each other. They more likely strike the walls of the chamber.

Now that we know something about the mean free paths of molecules, we can address the situation in which two connected chambers maintained at different temperatures will have unequal pressures at steady state.

$$\frac{P_1}{P_2} = \left(\frac{T_1}{T_2}\right)^{1/2} \tag{2.34}$$

In Section 2.12 we derived Eq. 2.34 from the Langmuir equation (Eq. 2.27). Flow of gas molecules through an orifice is governed by the Langmuir equation when the mean free path of the molecules is larger than the diameter of the orifice, that is, when $\lambda/a > 1$. This condition is called *Knudsen flow,* or *molecular flow.* The molecules pass through the orifice singly, without cooperative action. When the ratio of the mean free path to the orifice diameter is less than 0.01 ($\lambda/a < 0.01$), molecules passing through the orifice interact. This is called *viscous flow.* The Langmuir equation does not pertain to viscous flow. In the preceding section, we observed that opening a door between the cold exterior of a house and the warm interior does not produce a temperature difference between the two. At atmospheric pressure, the mean free path of molecules is about 10^{-7} m. A doorway is about one meter wide. The value of λ/a is about 10^{-7}, clearly less than 0.01. Hence the flow is viscous and Eq. 2.27 does not apply. The region between primarily molecular flow and primarily viscous flow ($1 < \lambda/a < 0.01$) is a transition region in which neither type of flow dominates.

2.14   DIFFUSION IN GASES

If in a mixture of gases at constant temperature, the concentration of the components is not uniform throughout, the gas molecules will move among themselves in such a way as to make the composition uniform after a long period of time. This process of movement is called *diffusion.* From empirical observation, the rate at which this movement takes place is linearly proportional to the negative of the concentration gradients (Fick's law). In mathematical terms:

$$J_m = -D \left(\frac{\partial C}{\partial x}\right)_T \tag{2.39}$$

where J_m is the flux of molecules through a particular plane. The flux has dimensions of moles per unit area per unit time. The units generally used are moles per square centimeter per second (mol cm^{-2} s^{-1}).

The term $(\partial C/\partial x)_T$ is the concentration gradient. Concentration is measured in moles per unit volume (length cubed), and the distance, x, is given in terms of length. The concentration gradient then has the units mass over the fourth power of length.

The units of concentration gradient generally used are moles per cubic centimeter per centimeter (mol cm^{-3} cm^{-1}).

Combining all these yields the units of D, the diffusion coefficient, as length squared over time. The units generally used are square centimeters per second (cm^2/s). If SI units are used throughout, the units of D are square meters per second (m^2/s).

One would expect this diffusion coefficient, D, to depend on the mean free path of molecules and the speed of the individual molecules. Because these quantities, average velocity and mean free path, are different for different atoms, it is difficult to calculate the interdiffusion coefficient of two molecular species (Refs. 5, 6). A simpler case, but one that is quite illuminating, is the calculation of a diffusion coefficient for two gases that have approximately the same chemical characteristics (e.g., two isotopes of the same element). This coefficient, called the *self-diffusion coefficient,* can be conveniently measured by studying the diffusion of a radioactive element into a nonradioactive isotope of the same element.

To derive the relationship between D and other properties of gaseous molecules, we refer to Figure 2.5. The horizontal tube in that figure contains a gas and a tracer of one of its radioactive isotopes. The concentration of the gas at $x = 0$ is C_0. At a location an infinitesimal distance from $x = 0$, the concentration is:

$$C = C_0 + \left(\frac{\partial C}{\partial x}\right) dx$$

Assume that the concentration of the radioactive isotope of the same gas is C^*. We can then write:

$$C^* = C_0^* + \left(\frac{\partial C^*}{\partial x}\right) dx$$

The flow of these radioactive molecules through the plane at $x = 0$ is related to the average absolute velocity of the molecules and their mean free path. In this

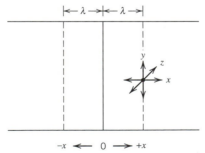

Figure 2.5 Movement of gas molecules in terms of mean free path (λ).

simplified calculation, we assume that all the radioactive molecules within a distance of a mean free path to the right of the plane zero have a probability moving through that plane before making another collision at a rate equal to:

$$N = \tfrac{1}{6} \langle v \rangle \, C^*_{x+\lambda}$$

To check on units, the average absolute velocity, $\langle v \rangle$, is measured in centimeters per second and the concentration in moles per cubic centimeter. The product of the two has the dimensions moles per square centimeter per second. The term $\tfrac{1}{6}$ arises because the atoms are moving in any one of six directions. Only one of these directions will carry a molecule through the plane zero. (Recognize that there are geometrical constraints to be observed. This is a simplified calculation. A good discussion of diffusion in gases, including a survey of more rigorous derivations, is in Ref. 6). Continuing with the calculation, the motion of atoms from the segment to the right of the zero plane can be expressed as follows:

$$\overleftarrow{N} = \tfrac{1}{6} \langle v \rangle \left[C^* + \left(\frac{\partial C^*}{\partial x} \right)_T (x + \lambda) \right]$$

The motion from the section to the left of plane zero can similarly be written as

$$\overrightarrow{N} = \tfrac{1}{6} \langle v \rangle \left[C^* + \left(\frac{\partial C^*}{\partial x} \right)_T (x - \lambda) \right]$$

The net motion is the difference between these two:

$$J = \overrightarrow{N} - \overleftarrow{N} = -\tfrac{1}{3} \langle v \rangle \left(\frac{\partial C^*}{\partial x} \right)_T \tag{2.40}$$

Comparing Eqs. 2.40 and 2.39 yields the diffusion coefficient of the gas:

$$D = \tfrac{1}{3} \langle v \rangle \lambda \tag{2.41}$$

It is known that the average absolute velocity of a molecule in an ideal gas is:

$$\langle v \rangle = \left(\frac{8kT}{\pi m} \right)^{1/2} \tag{2.21}$$

The mean free path is:

$$\lambda = \frac{kT}{\sqrt{2}\,\pi\sigma^2 P} \tag{2.38}$$

Therefore the diffusion coefficient is:

$$D = \frac{2}{3}\left(\frac{k}{\pi}\right)^{3/2}\frac{1}{m^{1/2}}\frac{T^{3/2}}{\sigma^2 P} \tag{2.42}$$

The diffusion coefficient in a gas is a function of the $\frac{3}{2}$ power of the temperature, the inverse first power of the pressure, and the inverse square root of the mass of the molecule. Increasing temperature increases diffusivity. Decreasing pressure increases diffusivity because the distance the molecules travel between collisions increases. Smaller, lighter molecules diffuse more rapidly, proportional to the inverse of the square root of their mass.

It is important to note that in this derivation, we assumed implicitly that the mean free path of the gaseous molecules is small in comparison to the container in which they are confined. If the mean free path is larger than the container dimensions, the molecules will collide with the wall more often than with themselves, and the concept of movement limited by intermolecular collisions is no longer valid.

An interesting illustration of the influence of mean free path on transport is provided by the comparison of the evaporation of tungsten in a vacuum and in an atmosphere of an inert gas, such as argon or nitrogen. For the purpose of this example, assume that a tungsten surface at 2200°C (2473 K) is separated by 3 cm from a cold surface. Any tungsten atoms that reach the cold surface are assumed to condense on it. If the space between the hot tungsten and the cold surface is a vacuum, then the mean free path of the tungsten atoms will be determined by the tungsten vapor pressure at 2473 K, which is 1.23×10^{-10} atm. At this pressure, the mean free path of the tungsten atoms will be about 10^5 cm, very much larger than the distance between the surfaces. Hence the rate of transport of tungsten to the cold surface will be determined by the maximum rate of evaporation of tungsten at 2473 K, calculated using the Langmuir equation (Eq. 2.31). The rate is calculated as follows:

$$J_{\mathrm{w}} = \frac{N^*}{A\tau} = \frac{P}{(2\pi mkT)^{1/2}}$$

$$J_{\mathrm{w}} = \frac{(1.23 \times 10^{-10})(1.013 \times 10^5)}{\left(2\pi\,\dfrac{184 \times 10^{-3}}{6.022 \times 10^{23}} \times 1.38 \times 10^{-23} \times 2473\right)^{1/2}}$$

$$J_{\mathrm{w}} = 4.85 \times 10^{16} \text{ atoms/(m}^2\cdot\text{s)}$$

Now suppose that the space between the hot and cold surfaces is filled with nitrogen at a pressure of 0.10 atm. At this pressure, the mean free path of the gaseous molecules is about 10^{-4} cm, a dimension very much smaller that the distance between the surfaces. In that case, the flow of tungsten atoms between the surfaces is limited by diffusion. For the purpose of this example, let us assume that the diffusion coefficient of tungsten atoms in nitrogen can be approximated by the self-diffusion coefficient of nitrogen and that we may calculate the latter by assuming an average gas temperature of 1400 K. Using Eq. 2.42, we find the diffusion coefficient:

$$D = \frac{2}{3} \left(\frac{1.38 \times 10^{-23}}{\pi} \right)^{3/2} \frac{(1400)^{3/2}}{\left(\frac{28 \times 10^{-3}}{6.022 \times 10^{23}} \right)^{1/2}} \frac{1}{(3 \times 10^{-10})^2 (1.013 \times 10^4)}$$

$$D = 1.66 \times 10^{-3} \text{ m}^2/\text{s}$$

The rate of transport of tungsten atoms through the nitrogen is given by

$$J_\text{W} = -D \frac{\Delta C}{\Delta x}$$

where ΔC is the difference in concentration of tungsten atoms between the two surfaces, and Δx is the distance between them.

At the cold surface, the concentration of tungsten atoms in the nitrogen vapor is zero, because all the tungsten atoms condense. The tungsten pressure at the hot surface is taken to be its vapor pressure. The concentration of tungsten in the vapor at the hot surface is:

$$C = \frac{N}{V} = \frac{P}{kT} = \frac{1.25 \times 10^{-5}}{(1.38 \times 10^{-23})(2473)} = 3.6 \times 10^{14} \text{ atoms/m}^3$$

The rate of transport of tunsten atoms through the nitrogen is then:

$$J_\text{W} = -(1.66 \times 10^{-3}) \left(\frac{3.6 \times 10^{14}}{-3 \times 10^{-2}} \right) = 2 \times 10^{13} \text{ atoms/(m}^2\cdot\text{s)}$$

By comparing the two rates of tungsten transport, one in a vacuum and the other in a nitrogen atmosphere, it is clear that the evaporation of tungsten is diminished considerably by the presence of nitrogen at a pressure of 0.1 atm.[4] At that pressure the resistance to flow provided by the nitrogen is clearly much greater than the resistance to flow provided by the limit on evaporation rate (Langmuir equation). To study the situation at lower nitrogen pressures, we can consider that the effect of two resistances to the flow of matter are analogous to the effect of two electrical resistances in series to the flow of electricity: that is, their resistive effects are additive. Appendix 2A provides the background for this approach.

---

[4]This is why incandescent lamps are filled with gas. If incandescent lamps with tungsten filaments were operated with an evacuated bulb, tungsten would evaporate quite rapidly. The nitrogen, or nitrogen–argon, atmospheres used in lamps prolongs their useful life.

The flow of tungsten atoms (J_W) based on the Langmuir equation (Eq. 2.31) is:

$$J_W = \frac{N_W^*}{A\tau} = \frac{P_W}{(2\pi mkT_W)^{1/2}}$$

where P_W is the tungsten pressure at temperature T_W.

The flow of tungsten atoms based on the diffusion equation is:

$$J_W = -D\frac{\Delta C}{\Delta x} = -\frac{D}{kT_W\Delta x}(P_W)$$

In terms of driving force (P_W) and resistance, this equation can be written as follows

$$J_W = \frac{P_W}{kT_W\Delta x/D}$$

The resistance term is $RT_W\Delta x/D$. Note that the sign of the equation has been changed to reflect the treatment as an electrical resistance.

Considering both resistances, the flow of tungsten atoms is

$$J_W = \frac{P_W}{(2\pi mkT_W)^{1/2} + kT\Delta x/D} \qquad (2.43)$$

where D is the diffusion coefficient of tungsten atoms in the medium through which it is diffusing. The self-diffusion coefficient of that medium is

$$D = \frac{2}{3}\left(\frac{k}{\pi}\right)^{3/2}\frac{1}{m^{1/2}}\frac{T^{3/2}}{\sigma^2 P} \qquad (2.42)$$

where T is the temperature of the medium and P is its pressure.

Based on Eq. 2.42, as the pressure of the gas in the diffusion medium (nitrogen in the example) decreases, the diffusivity D increases. As the diffusivity increases, the diffusion resistance term in Eq. 2.43 becomes less significant. When the mean free path of the diffusing atom becomes large as compared to the size of the diffusion distance, the diffusivity effectively becomes infinite, and Eq. 2.42 becomes equivalent to the Langmuir equation (Eq. 2.31).

2.15   ELASTICITY OF RUBBER

The deformation characteristics of polymers are different from those of metals and ceramics. In general, the elastic moduli of metals and ceramics are much larger than those of polymers. In addition, the elastic deformations of metals and ceramics dur-

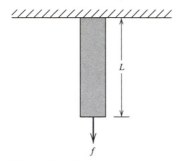

Figure 2.6 Elongation of a rubber rod.

ing tensile extensions are very much lower than those of rubberlike polymers (elastomers). Elastic extensions of metals and ceramics seldom exceed 1%. Elastomers can have elastic (reversible) extensions exceeding 500 or 600%.

The elastic behavior of metals and ceramics is, by comparison to polymers, easy to understand. During elastic deformation, the atoms in metals and ceramics remain in the same position relative to one another, and the forces required to deform the material can be estimated from a knowledge of the interatomic forces. In polymers, by contrast, deformation of the material requires a deformation of the individual molecules, an uncoiling of the long chains that form the polymer. Through the application of statistical thermodynamics, a model of the mechanical behavior of rubberlike polymeric materials (elastomers) has been created based on entropy changes in the solid during extension.[5]

To lay the foundation for our study of rubber elasticity, let us first consider the elongation of a rubber rod from the macroscopic point of view (Figure 2.6). If we neglect pressure–volume effects, the expression for dU, the change of internal energy of the rod, is

$$dU = T\,dS + f\,dL \qquad (2.44)$$

where f is the force applied and L is the length of the sample.

From the definition of the Helmholtz free energy, $F = U - TS$:

$$dF = -S\,dT + f\,dL \qquad (2.45)$$

The Maxwell relation resulting from Eq. 2.45 is:

$$-\left(\frac{\partial S}{\partial L}\right)_T = \left(\frac{\partial f}{\partial T}\right)_L \qquad (2.46)$$

---

[5]Excellent reviews of this subject are included in Ref. 7 (Flory) and in Ref. 8 (Hiemenz). The presentation in this chapter will follow the general order in Ref. 8.

From Eq. 2.44:

$$f = \left(\frac{\partial U}{\partial L}\right)_T - T\left(\frac{\partial S}{\partial L}\right)_T \qquad\qquad (2.47)$$

Combining Eqs. 2.46 and 2.47:

$$f = \left(\frac{\partial U}{\partial L}\right)_T + T\left(\frac{\partial f}{\partial T}\right)_L \qquad\qquad (2.48)$$

A typical force–length (f–L) relationship for rubber at two different temperatures, T_1 and T_2, is shown schematically in Figure 2.7.

An ideal rubber is defined as one for which

$$\left(\frac{\partial f}{\partial T}\right)_L = \frac{f}{T} \qquad\qquad (2.49)$$

Combining Eqs. 2.49 and 2.48 yields

$$\left(\frac{\partial U}{\partial L}\right)_T = 0 \qquad\qquad (2.50)$$

This means that the internal energy of an ideal rubber does not change with length at constant temperature. An adiabatic stretching process constitutes an input of energy into the rubber rod. That energy input must manifest itself in some way. If

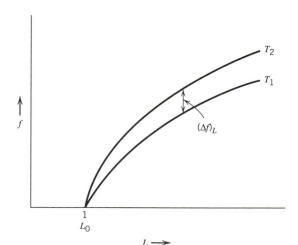

Figure 2.7 The force-versus-length relationship for an ideal rubber at two temperatures.

the internal energy (at constant temperature) does not increase as the length increases, we must conclude that the temperature of the sample does. This can be demonstrated qualitatively by sensing the temperature change of a rubber band as it is elongated. To perform a simple demonstration, quickly stretch a rubber band to two or three times its normal length. Temperature, for this demonstration, can be sensed by pressing the rubber band against one's upper lip. The result is reversible; that is, the rubber band cools upon shrinking to its normal, unstressed length. (See Vol. I, Section 3.12.)

Combining Eqs. 2.50 and 2.47 yields

$$f = -T \left(\frac{\partial S}{\partial L}\right)_T \tag{2.51}$$

Equation 2.51 tells us that the force required to stretch rubber isothermally is related to the entropy change of the rubber with length. The internal energy of the rubber (ideal) does not change as it is elongated isothermally, but its entropy does. This implies that we are not stretching the polymer (rubber) molecules themselves during the process, but we are changing the degree to which they are aligned. The greater their degree of alignment, the lower their entropy. The rest of this section is devoted to calculating the entropy of a polymer (rubber) as a function of its length, and the related degree of alignment of its molecules.

To explore this relationship quantitatively will require a digression into the statistics of the random walk. The "random walk" refers to the position of a particle that undergoes random steps of equal length in various directions. To illustrate, let us consider the random walk in one dimension, the x direction. The particle moves in discrete jumps of equal length to the right and to the left.

To guide the motion of the particle, consider the tossing of a fair coin. If head turns up, we will move one step to the right; if tail turns up, we will move one step to the left. From the binomial probability distribution, the probability that the particle makes N_R moves to the right out of a total of N moves will be:

$$P(N_R, N) = \frac{N!}{N_R! N_L!} P_R^{N_R} P_L^{N_L} \tag{2.52}$$

where P_R and P_L are the probabilities of a right and left movement, respectively. Each of the two terms will be $\frac{1}{2}$ for an unbiased coin.

With moves to the right being considered positive (in an algebraic sense), the displacement, x, after N steps of length l will be:

$$x = (N_R - N_L)l$$

We note that

$$N = N_R + N_L$$

and then

$$N_R = \frac{1}{2}\left(N + \frac{x}{l}\right) \quad \text{and} \quad N_L = \frac{1}{2}\left(N - \frac{x}{l}\right)$$

For a fair coin, the probability of a head is one-half, and the probability of a tail is the same; thus Eq. 2.52 becomes:

$$P(x, N) = \frac{N!}{\left(\dfrac{N + x/l}{2}\right)! \left(\dfrac{N - x/l}{2}\right)!}\left(\frac{1}{2}\right)^N \qquad (2.53)$$

Applying the Stirling approximation ($\ln N! = N \ln N - N$) and some algebra yields:

$$-\ln P(x, N) = \left(\frac{Nl + x}{2l}\right)\ln\left(1 + \frac{x}{Nl}\right) + \left(\frac{Nl - x}{2l}\right)\ln\left(1 - \frac{x}{Nl}\right)$$

Because the value of X/Nl is much less than one when N is large, we can approximate the value of the logarithmic terms as follows:

$$\ln\left(1 + \frac{x}{Nl}\right) = \frac{x}{Nl} - \frac{1}{2}\left(\frac{x}{Nl}\right)^2$$

Thus:

$$\ln P(x, N) = -\frac{x^2}{2Nl^2}$$

At this point we will solve for $P(x, N)$ and introduce a multiplying factor, K, which will be evaluated to set sum of the probabilities to 1. The derivation is given in Appendix 2B.

$$P(x, N) = K \exp\left(-\frac{x^2}{2Nl^2}\right) \qquad (2.54)$$

$$K \int_{-\infty}^{+\infty} \exp\left(-\frac{x^2}{2Nl^2}\right) dx = 1$$

$$K = (2\pi Nl^2)^{-1/2}$$

$$P(x, N) = (2\pi Nl^2)^{-1/2} \exp\left(-\frac{x^2}{2Nl^2}\right) \qquad (2.55)$$

Equation 2.55 gives the probability of finding the particle at a distance x from the origin after N moves, where each move is of length l. The mean squared average distance of the particle from the origin after N random moves can be calculated by evaluating the expression:

$$\overline{x^2} = \int_{-\infty}^{+\infty} x^2 P(x,\ N)\ = 2 \int_{0}^{+\infty} x^2 (2\pi N l^2)^{-1/2} \exp\left(-\frac{x^2}{2Nl^2}\right) dx \qquad (2.56)$$

The evaluation of the integral is given in Appendix 2C with the following result[6]:

$$\overline{x^2} = Nl^2 \qquad (2.57)$$

If instead of moving in one direction, we had moved in three orthogonal directions, the probability of finding the particle in a cubical volume of dimensions dx, dy, dz at a position x,y,z after N moves would be given by[7]:

$$P(x,y,z,N)\ dx\ dy\ dz = P\left(x, \frac{N}{3}\right) P\left(y, \frac{N}{3}\right) P\left(z, \frac{N}{3}\right) dx\ dy\ dz$$

$$\qquad (2.58)$$

$$P(x,y,z,N)\ dx\ dy\ dz = \left(2\pi \frac{N}{3} l^2\right)^{-3/2} \exp\left(-\frac{3(x^2 + y^2 + z^2)}{2Nl^2}\right) dx\ dy\ dz$$

Equation 2.58 also can be applied to the probability of finding the end of a polymeric molecule at any position x,y,z starting from the origin ($x = 0$, $y = 0$, $z = 0$). Assume that the molecule, containing N mers, each of length l, is free to

---

[6]The model described above suggests a diffusion process in which particles move by random jumps. If we consider diffusion of a radioactive tracer of a material placed on the interface between two rods, the concentration of the radioactive tracer as a function of time and distance from the interface is given by:

$$\frac{C(x,\ t)}{C_0} = \frac{1}{2\sqrt{\pi Dt}} \exp\left(-\frac{x^2}{4Dt}\right)$$

Comparing the equation above with Eq. 2.47 yields

$$2Nl^2 = 4Dt$$

$$D = \frac{1}{2}\left(\frac{N}{t}\right) l^2$$

The diffusion coefficient, D, is then the product of the jump frequency $N/t = \Gamma$ and the square of the jump distance. (This relationship will be reviewed in Chapter 5.)

[7]The value N is changed to $N/3$ because each move can take place in one of three directions.

bend in any direction at any time (including bending back on itself, which is, admittedly, a physical impossibility). To establish the correspondence with the random walk model, recognize that each mer is a "jump." Asking where the end of the polymer chain containing N mers is located is the same as asking where a particle undergoing a random walk is located after N jumps.

Equation 2.58 can be changed to radial coordinates as follows:

$$x^2 + y^2 + z^2 = R^2$$

$$dx\, dy\, dz = 4\pi R^2\, dR$$

$$P(R, N)\, dR = \left(2\pi \frac{N}{3} l^2\right)^{-3/2} 4\pi R^2 \exp\left(-\frac{3R^2}{2Nl^2}\right) dR \tag{2.59}$$

The mean square value of R, the distance from the origin to the end of the molecule, can be calculated as follows:

$$\overline{R^2} = \int_0^\infty P(R, N) R^2\, dR = 4\pi \left(\frac{2\pi Nl^2}{3}\right)^{-3/2} \int_0^\infty R^4 \exp\left(-\frac{3R^2}{2Nl^2}\right) dR$$

This integral can be evaluated, resulting in

$$\overline{R^2} = Nl^2 \qquad \text{or} \qquad l = \frac{R_{rms}}{N^{1/2}} \tag{2.60}$$

It follows from the foregoing that the root-mean-square distance of the end of a molecule from the origin is a function of the length of the mer, l, and the square root of the number of mers in a chain, N.

With all this background, let us turn now to the question of deformation. When a solid is deformed, its length is increased by a value ΔL:

$$L = L_0 + \Delta L$$

We can define α as the ratio of the length to its original length:

$$\alpha = \frac{L}{L_0} \tag{2.61}$$

The value for elongation normally used is $\varepsilon = \Delta L/L_0$:

$$\alpha - 1 = \varepsilon \tag{2.62}$$

When a solid with original dimensions x_0, y_0, and z_0 is stretched in the z direction at *constant volume,* the dimensions of the solid will change as follows:

$$z_0 \rightarrow \alpha z_0$$

$$x_0 \rightarrow \frac{1}{\alpha^{1/2}} x_0$$

$$y_0 \rightarrow \frac{1}{\alpha^{1/2}} y_0$$

The volume will be, as before, equal to x_0, y_0, z_0.

$$V = \left(\frac{1}{\alpha^{1/2}}\right) x_0 \left(\frac{1}{\alpha^{1/2}}\right) y_0 \alpha z_0 = x_0 \, y_0 \, z_0 \qquad (2.63)$$

When the solid is stretched, the probability distribution of the ends of the molecules changes. No longer is the location of the end of a molecule independent of direction. If it is stretched in the z direction, a greater concentration of ends will be found pointed in that direction. This is because a cubical volume of an elastomer becomes, upon stretching, a rectangular parallelopiped elongated in the z direction. The x–y plane (facing the z direction) is reduced in size, which increases the concentration of molecule ends on that face. The opposite is true of the x–z planes and the y–z planes (Figure 2.8).

The probability function for the unstretched polymer P_u represents the probability

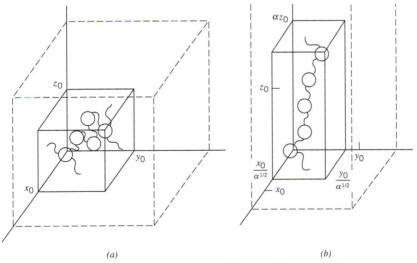

(a)   *(b)*

Figure 2.8 The deformation of a polymer subchain. (*a*) The original coordinates of the end of the subchain are x_0, y_0, and z_0. (*b*) The same coordinates as in (a) are $x_0/\alpha^{1/2}$, $y_0/\alpha^{1/2}$, and αz_0.
Source: Ref. 8.

that the end of a molecule with N mers, each of length l, will be found in a volume $dx\,dy\,dz$ at a distance x, y, z from the origin.

$$P_u dx\,dy\,dz = (\tfrac{2}{3}\pi N l^2)^{-3/2} \exp\left[-\frac{3(x^2 + y^2 + z^2)}{2Nl^2}\right] dx\,dy\,dz \qquad (2.64)$$

where N_c is the number of segments (mers) and l_0 is the length of the segment.
This same probability function after stretching, P_s, is

$$P_s dx\,dy\,dz = (\tfrac{2}{3}\pi N l^2)^{-3/2} \exp\left[-\frac{3(x^2/\alpha + y^2/\alpha + z^2\alpha^2)}{2Nl^2}\right] \qquad (2.65)$$

We can relate these changes in probability to a change in entropy using the Boltzmann formulation (Eq. 2.3) because the energy of the system was assumed not to change during the elongation (see Eq. 2.50). Remember, that the elongation forces in the case of elastomers are being related to an entropy effect. The evaluation of the values for Ω_s and Ω_u involves the integration of the probability function over all the possible values of x, y, and z for the stretched and unstretched cases respectively (see Ref. 7, pp. 465–468). The entropy change upon stretching is

$$\Delta S = S_s - S_0 = k(\ln \Omega_s - \ln \Omega_u) \qquad (2.66)$$

This change in entropy refers to the entropy change of a typical polymer chain, a chain labeled i, for example:

$$\Delta S_i = -3k\,\frac{\left[\left(\dfrac{1}{\alpha} - 1\right)x^2 + \left(\dfrac{1}{\alpha} - 1\right)y^2 + (\alpha^2 - 1)z^2\right]}{2Nl^2} \qquad (2.67)$$

Remembering that the average squared distance from the origin is:

$$x^2 + y^2 + z^2 = Nl^2$$

and that in the unstretched case

$$x^2 = y^2 = z^2 = \tfrac{1}{3}Nl^2$$

This transforms Eq. 2.67 into

$$\Delta S_i = -\frac{3k}{2}\left[\frac{2}{3}\left(\frac{1}{\alpha} - 1\right) + \frac{1}{3}\alpha^2 - \frac{1}{3}\right]$$

$$\Delta S_i = -\frac{k}{2}\left[\alpha^2 + \frac{2}{\alpha} - 3\right] \qquad (2.68)$$

The total entropy change of the sample is the product of the number of chains, v, multiplied by the average entropy change of each chain (molecule).

$$\Delta S = -\frac{kv}{2}\left(\alpha^2 + \frac{2}{\alpha} - 3\right) \tag{2.69}$$

where v is the number of polymer chains in the sample.

Referring now to the macroscopic thermodynamics of the rubber rod being stretched, we use Eq. 2.47 and Eq. 2.50 to write

$$f = -T\left(\frac{\partial S}{\partial L}\right)_T = -\frac{T}{L_0}\left(\frac{\partial S}{\partial \alpha}\right)_T$$

Note that $L = \alpha L_0$

$$f = \frac{kTv}{L_0}\left(\alpha - \frac{2}{\alpha^2}\right) \tag{2.70}$$

Stress is force per unit area

$$\sigma = \frac{f}{A} = \frac{kTv}{AL_0}\left(\alpha - \frac{2}{\alpha^2}\right) = \frac{kTv}{V}\left(\alpha - \frac{2}{\alpha^2}\right) \tag{2.71}$$

Note that the value of stress changes sign at the value $\alpha = 1$, as it should. (At $L = L_0$, $\alpha = 1$.)

$$\sigma = \frac{kTv}{V}\left[\frac{L}{L_0} - \left(\frac{L_0}{L}\right)^2\right] \tag{2.72}$$

Differentiating Eq. 2.72 yields:

$$d\sigma = \frac{kTv}{V}[L_0^{-1} - 2L_0^2L^{-3}]dL$$

In terms of α, we have

$$d\sigma = \frac{kTv}{V}\left(\alpha + \frac{2}{\alpha^2}\right)\frac{dL}{L} \tag{2.73}$$

This relationship between stress and strain is the Young's modulus:

$$d\sigma = E\frac{dL}{L}$$

and $v/\underline{V} = \rho N_A/M_c$, where N_A is Avogadro's number and $\rho \underline{V} = M_c$, the molecular weight.

$$E = \frac{RT\rho}{M_c}\left(\alpha + \frac{2}{\alpha^2}\right) \tag{2.74}$$

Several things about Eq. 2.74 are interesting. First, the modulus increases with temperature. This is in contrast to the moduli of metals or ceramics, which decrease with increasing temperature. It is also interesting to note that the modulus increases as the value M_c, the polymer chain length, decreases. The value of M_c can be decreased by reducing the effective length of the polymer molecules through cross-linking of the chains.

Moreover, in contrast to metals and ceramics, the modulus is not constant in the elastic behavior region. The modulus of rubber depends on the amount of deformation. For values of α very much greater than 1, the first term in Eq. 2.74 predominates, and the modulus is directly proportional to length, or α. For values of α closer to 1 ($L/L_0 = 2$), the second term is more significant. At very low deformations, the modulus is approximately:

$$E = \frac{3RT\rho}{M_c} \quad \text{for} \quad \alpha \approx 1 \tag{2.75}$$

A word of caution is in order. The elastic behavior of rubber is shown schematically in Figure 2.9. The model used to derive the stress–elongation behavior is valid when the elasticity is controlled by an entropy effect, that is, when the individual

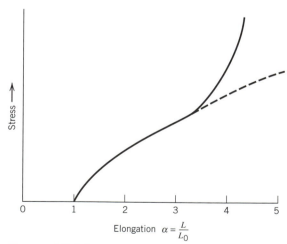

Figure 2.9 Schematic stress–elongation curve for rubber.

molecules are being oriented in the direction of the applied stress. After a certain level of elongation, the molecules will be largely oriented in the stress direction. At this point, increasing stress will tend to elongate the molecules themselves. The stress required to do so is larger than that required to orient the molecules, and the stress curve deviates from the one derived in this section. The extrapolation of the stress–elongation behavior described in Eq. 2.71 is shown as a dashed line.

REFERENCES

1. Chandler, David, *Introduction to Modern Statistical Mechanics,* Oxford University Press, New York, 1987.
2. Hill, Terrell L., *An Introduction to Statistical Thermodynamics,* Dover, New York, 1986.
3. Mayer, J. E., and Mayer, M. G., *Statistical Thermodynamics,* Wiley, New York, 1959.
4. Dole, Malcolm, *An Introduction to Statistical Thermodynamics,* Prentice-Hall, Englewood Cliffs, NJ, 1954.
5. Castellan, Gilbert W., *Physical Chemistry* 3rd ed, Benjamin/Cummings, New York, 1983 (especially Chapter 29).
6. Jost, W., *Diffusion in Solids, Liquids, and Gases,* Academic Press, New York, 1960 (especially Chapter X).
7. Flory, Paul J., *Principles of Polymer Chemistry,* Cornell University Press, Ithaca, NY, 1960.
8. Hiemenz, Paul C., *Polymer Chemistry,* Dekker, New York, 1984.

APPENDIX 2A

Heat Flow (or Diffusion) Through a Two-Layered Wall

In Section 2.14, it was asserted that the multiple resistances to flow can be treated as if they were series resistances in an electrical circuit, that is, the total resistance is the sum of the resistances. An analysis of a simple heat flow example will make this plausible. The equation governing heat flow (Fourier's law) is similar in form to Fick's law governing mass flow:

$$J_Q = -k \frac{\Delta T}{\Delta x} \tag{2A.1}$$

where k is the thermal conductivity, and ΔT is the temperature difference across the distance Δx.

Consider the flow of heat through a composite wall consisting of two layers. The temperatures and dimensions are given in Figure 2A.1. At steady state the heat flow through the sections 1 and 2 will be equal, hence:

$$J_Q = -k_1 \frac{T_A - T_B}{\Delta x_1} = -k_2 \frac{T_B - T_C}{\Delta x_2} \tag{2A.2}$$

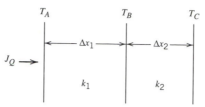

Figure 2A.1 Temperatures and dimensions for two-layered wall.

Rearranging to treat the driving force as ΔT and the resistance to flow as $R = \Delta x/k$:

$$J_Q = -\frac{T_A - T_B}{\Delta x_1/k_1} = -\frac{T_B - T_C}{\Delta x_2/k_2} \tag{2A.3}$$

$$J_Q = -\frac{T_A - T_B}{R_1} = -\frac{T_B - T_C}{R_2} \tag{2A.4}$$

Solving for T_B:

$$T_B = J_Q R_1 + T_A \tag{2A.5}$$

Substituting in Eq. 2A.4:

$$J_Q = -\frac{(J_Q R_1 + T_A) - T_C}{R_2}$$

$$J_Q R_2 = -J_Q R_1 - T_A + T_C$$

$$J_Q = -\frac{T_A - T_C}{R_1 + R_2} = -\frac{T_A - T_C}{\Delta x_1/k_1 + \Delta x_2/k_2}$$

Thus the flow through the composite wall may be treated as if it were a series electrical circuit. The resistances to flow may be added. This example used two sections. The same result is obtained with more than two sections.

APPENDIX 2B

Evaluation of K: Equation 2.54

$$P(x_1, N) = K \exp\left(-\frac{x^2}{2Nl^2}\right)$$

$$K \int_{-\infty}^{\infty} \exp\left(-\frac{x^2}{2Nl^2}\right) dx = 1$$

$$2K \int_0^{\infty} \exp\left(-\frac{x^2}{2Nl^2}\right) dx = 1$$

Definite integral:

$$\int_0^\infty \exp\left(-\frac{x^2}{2Nl^2}\right) dx = \int_0^\infty \exp(-a^2 x^2) dx$$

$$= \frac{1}{2a} \sqrt{\pi}$$

$$a^2 = \frac{1}{2Nl^2} \; ; \quad a = \left(\frac{1}{2Nl^2}\right)^{1/2}$$

$$\int_0^\infty \exp - \left(\frac{x^2}{2Nl^2}\right) dx = \frac{(2Nl^2\pi)^{1/2}}{2}$$

$$\frac{2K}{2} (2Nl^2\pi)^{1/2} = 1$$

$$K = (2\pi Nl^2)^{-1/2}$$

APPENDIX 2C

Calculation of $\bar{x}^2$

$$\bar{x}^2 = \int_{-\infty}^\infty x^2 P(x,N) = 2 \int_0^\infty x^2 (2\pi Nl_0^2)^{-1/2} \exp\left(-\frac{x^2}{2Nl^2}\right) dx$$

$$= 2(2\pi Nl^2)^{-1/2} \int_0^\infty x^2 \exp\left(-\frac{x^2}{2Nl_2}\right) dx$$

Definite integral

$$\int_0^\infty x^2 \exp(-bx^2) dx = \frac{1}{4b} \left(\frac{\pi}{b}\right)^{1/2}$$

$$b = \frac{1}{2Nl^2}$$

$$\bar{x}^2 = 2[2\pi Nl^2]^{-1/2} \left[\frac{1}{4(1/2Nl^2)} \left(\frac{\pi}{1/2Nl^2}\right)^{1/2}\right]$$

$$\bar{x}^2 = 2[2\pi Nl^2]^{-1/2} \left[\frac{Nl^2}{2}\right] [2\pi Nl^2]^{1/2}$$

$$\bar{x}^2 = Nl^2$$

PROBLEMS

2.1 To determine the vapor pressure of magnesium, shavings of the metal are placed in an inert, nonvolatile container with a small hole in its top. The container is suspended in a vacuum furnace at 700 K, and its weight loss rate is measured.

    (a) Based on the information given below, calculate the vapor pressure of magnesium at 700 K.

    (b) If the estimated error in the measurement of the hole diameter is $\pm 1\%$, and the estimated error in the weight loss measurement is $\pm 3\%$, what error limits should be placed on the pressure calculated from these data?

DATA

Diameter of circular hole in container = 1 mm
Weight loss rate = 2.074 mg/h
Molecular weight of Mg = 24.3 g/mol

2.2 Liquid cadmium at its melting point is being held in a furnace under pure nitrogen. Cadmium atoms evaporate from the surface of the liquid and diffuse through the nitrogen to the top of the furnace 10 cm away, where they are deposited.

    Calculate the rate of transport of cadmium (in grams per square centimeter of surface per second) as a function of nitrogen pressure. Prepare a rough plot of the transport rate as a function of nitrogen pressure from one atmosphere to 10^{-9} atm. Recognize that the transport of cadmium may be limited by either the rate of diffusion through the nitrogen or the evaporation from the surface.

DATA

For Cd in nitrogen, $D = 0.17$ cm²/s when the nitrogen pressure is 1 atm at the melting point of Cd (594 K).
At 594 K the vapor pressure of Cd is 100 μm Hg
(1 μm Hg = 10^{-6} m Hg = 1.32×10^{-6} atm).

2.3 Quantum dots are very fine particles of materials such as high purity silicon. We are interested in keeping the "dots" clean. Estimate how many molecules of gas (oxygen) are adsorbed on the surface of one of the "dots" specified in the next paragraph.

    The "dot" in question is a sphere with a diameter of 1000 Å. Its residence time in a vacuum apparatus is 10 ms. The pressure in the apparatus is 10^{-10} atm, and the temperature is about 300 K. Assume that all the gas molecules in the apparatus are oxygen, and that all the molecules that strike the surface of the dot stick to it.

2.4 The accompanying diagram shows schematically an apparatus for depositing silver on samples in a vacuum chamber. The silver is evaporated from a source with an area of 1 cm² and is held at a temperature of 1600 K. The sample to be coated is about 30 cm from the source. Assume that the silver atoms evaporate uniformly in all directions.

(a) What is the maximum rate of deposition of silver on the sample, in grams per square centimeter per second?

(b) If the background pressure (the pressure of oxygen before and during the deposition) in the vacuum furnace is 10^{-9} atm, what will be the maximum deposition rate of oxygen on the sample, in grams per square centimeter per second? Assume that the average gas (oxygen) temperature near the deposition surface is 700 K.

(c) What will be the impurity level of oxygen in the silver deposited (in parts per million by weight)?

DATA

At 1600 K, the vapor pressure of silver is 10^{-3} atm.
The atomic weight of silver is 107.8 g/mol.
Silver vapor is monatomic.
The molecular weight of oxygen is 32 g/mol.

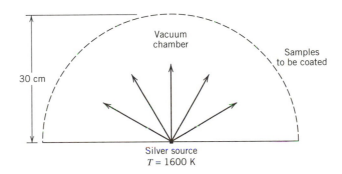

2.5 In a vacuum furnace (shown schematically in the accompanying diagram), an electronic device on the cold wall is being coated with aluminum. The aluminum is deposited from the vapor phase which is in equilibrium with liquid aluminum at 1200 K. You may assume that the mean free path of the aluminum atoms in the vapor is very much larger than the dimensions of the furnace.

(a) What is the deposition rate of aluminum (atoms per square meter per second) on the device, assuming that all the atoms that hit the sample stick to it?

(b) How long does it take for a 1 μm (10^{-6} m) layer to form on the device?

(c) If the temperature of the molten aluminum (nominally at 1200 K) can be controlled only to ± 5 K, what is the expected fractional variation ($\Delta l / l$) in the thickness of the aluminum deposited?

For aluminum:

$$\text{Vapor pressure: } \ln P \text{ (atm)} = 13.613 - \frac{37{,}334}{T \text{ (K)}}$$

$$\text{Atomic weight} = 27 \text{ g/mol}$$
$$\text{Density} = 2.7 \text{ g/cm}^3$$

Aluminum vapor is monatomic.

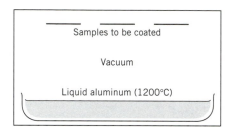

2.6 Coatings of elemental silicon are sometimes applied using chemical vapor decomposition (CVD). In this process gaseous silane (SiH_4) at a high temperature is carried by an inert gas over the surface to be coated. When a silane molecule reaches the hot surface, it decomposes into solid silicon and gaseous hydrogen:

$$SiH_4 \text{ (gas)} \rightarrow Si \text{ (solid)} + 2H_2 \text{ (gas)}$$

Under the operating conditions being considered, you may assume that the deposition rate is controlled by the diffusion of silane through a stagnant layer of inert gas 0.1 mm thick to the silicon surface.

The total pressure in the apparatus is one atmosphere. The pressure of silane in the inert gas is 0.1 atm. The silane concentration at the surface is assumed to be zero. The temperature of the apparatus is 900 K.

(a) What is the concentration of silane in the vapor away from the surface at 900 K *(in grams per cubic centimeter)*?

(b) What is the rate of deposition of silicon *(grams per square centimeter per second)* at 900 K?

(c) If all other conditions are held constant, what would be the effect of raising the temperature to 1000 K? If there is an effect, give a quantitative estimate of the magnitude.

For silicon:

Density of solid = 2.3 g/cm³

Atomic weight = 28 g/mol

The diffusion coefficient of silane in the inert gas at 900 K is approximately 0.15 cm²/s.

Chapter 3

Defects in Solids

There are many reasons for studying imperfections in solids. Important properties of solids depend on defects in the atomic and electronic structure of crystals. For example, it is well established that diffusion in metallic crystals depends on the presence of vacancies in the lattice. Plastic deformation in metals is known to be associated with the motion of dislocations. The electrical conductivity of many ionic solids is associated with the movements of vacancies or interstitial ions. The electrical behavior of semiconducting crystals is related to electronic imperfections. These are but a few of the practical reasons for discussing imperfections. However, it is not necessary to look only at practical considerations to justify the study of imperfections. Imperfections are interesting per se, in crystals, in oriental rugs, in postage stamps, and even in people. It has been said that a perfectly righteous person is to be respected and admired . . . but it is the sinner who makes the more intriguing study . . . and more interesting company.

This chapter treats crystal imperfections from the thermodynamic point of view. It has been noted that imperfections can be observed and analyzed in two extremes of concentration. When very dilute, imperfections are present in such low concentrations that they may be regarded as discrete entities with well-defined individual natures. This is true of point defects such as vacancies or interstitial atoms. Each such defect can be assigned properties, such as an energy and an entropy of formation, that are independent of the presence of all other imperfections. An individual defect may interact with others to form simple combinations such as pairs. Again, these simple combinations may be treated as separate entities and assigned energies and entropies of their own.

At the opposite extreme of concentration, the density of defects is so high that each defect loses its individual property to the nature of the group. Surfaces, such as grain boundaries, fall into this category. These group structures may be assigned properties such as interfacial tension, that make their description in thermodynamic terms possible.

This chapter deals with the thermodynamics of point defects. The thermodynamics of surfaces is treated in Chapter 4.

3.1   STRUCTURAL POINT DEFECTS IN ELEMENTAL CRYSTALS

A missing atom in the crystal structure of an elemental crystal is a *vacancy* or a *point defect.* If the absence of an atom on a lattice site causes no changes in the rest of the crystal, we can apply in a simple way the principles learned in Section 2.6 to the case of vacancies in elemental crystals. The crystal will consist of N atoms and n vacant lattice sites (vacancies). Assume that the energy to create such a vacancy is given by E_v. There are thus two states of a lattice site to be considered:

$$\text{an occupied state} \qquad E = 0$$
$$\text{an unoccupied state} \qquad E = E_v$$

The probability of finding such an unoccupied state is related to the energy required to produce the vacancy and the temperature of the crystal (from Section 2.5, Eq 2.7) is:

$$P_v = \frac{\exp(-E_v/kT)}{Z} \tag{3.1}$$

where $Z = 1 + \exp(-E_v/kT)$

$$P_v = \frac{n}{n + N} = \frac{\exp(-E_v/kT)}{1 + \exp(-E_v/kT)} \tag{3.2}$$

$$\frac{n}{N} = \exp\left(\frac{-E_v}{kT}\right)$$

Equation 3.2 shows that at temperatures above absolute zero, elemental crystals at equilibrium contain vacancies. A typical value, such as the one for elemental aluminum, is $E_v = 0.75$ eV (72.4 kJ/mol).

Actually, the absence of an atom on a lattice site changes the vibration patterns of neighboring atoms (Figure 3.1). We must therefore take into account not only the energy (or enthalpy) required to produce a vacancy, but also the entropy changes in

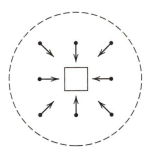

Figure 3.1 Effects of vacancy on nearest neighbors: all are ascribed to the vacancy.

the area surrounding that vacancy. At equilibrium, the Gibbs free energy of the crystal is at a minimum with respect to the number of vacancies present at constant temperature and pressure. The molar Gibbs free energy change, ΔG between the imperfect (G_i) and perfect crystal (G_p) is

$$\Delta G = G_i - G_p = n\,\Delta \underline{H}_v - T\,\Delta \underline{S} \qquad (3.3)$$

where $\Delta \underline{H}_v$ = enthalpy of formation of a vacancy
n = number of vacancies.

The entropy term, $\Delta \underline{S}$, can be expressed as follows:

$$\Delta \underline{S} = n\,\Delta \underline{S}_v + \Delta \underline{S}_c \qquad (3.4)$$

The total change in molar entropy $\Delta \underline{S}$ is equal to the sum of entropy change required to create the vacancy $\Delta \underline{S}_v$ and the configurational entropy $\Delta \underline{S}_c$, which is related to the uncertainty in the spatial location of the vacancy. Assuming a random distribution of vacancies, this configurational entropy term can be expressed as follows:

$$\Delta S_c = k \ln \Omega = k \ln \left(\frac{(N + n)!}{N!n!} \right) \qquad (3.5)$$

After applying the Stirling approximation, we can write an expression for $\Delta \underline{G}$:

$$\Delta \underline{G} = n\,(\Delta \underline{H}_v - T\,\Delta \underline{S}_v) - kT[(N + n)\ln(N + n) - N \ln N - n \ln n] \qquad (3.6)$$

The minimum of $\Delta \underline{G}$ as a function of the number of imperfections is:

$$\left(\frac{\partial \Delta \underline{G}}{\partial n} \right)_T = \Delta \underline{H}_v - T\,\Delta \underline{S}_v + kT \ln \left(\frac{n}{N + n} \right) = 0 \qquad (3.7)$$

Solving for the fraction of vacant sites (x_v):

$$x_v = \frac{n}{N + n} = \exp\left(\frac{\Delta \underline{S}_v}{k}\right) \exp\left(\frac{-\Delta \underline{H}_v}{kT}\right) \tag{3.8}$$

In general, for imperfections,

$$x_i = A \exp\left(\frac{-\Delta \underline{H}_i}{kT}\right) \tag{3.9}$$

where $A = \theta \exp(\Delta \underline{S}_v/k)$ and θ is a factor to account for variations in crystallography. Note that $x_i = n_i/N$ because $n_i \ll N$.

By taking the natural logarithm of both sides, Eq. 3.9 may also be written as follows:

$$\ln x_i = \ln A - \left(\frac{\Delta \underline{H}_i}{k}\right)\left(\frac{1}{T}\right) \tag{3.10}$$

In this form it is apparent that a graph of the natural logarithm of the defect concentration (or some quantity or property linearly dependent on it) on the ordinate, and inverse absolute temperature on the abscissa, will be a straight line with a slope equal to the negative of the enthalpy of formation of the imperfection divided by k, the Boltzmann constant (Figure 3.2).

From the foregoing analysis, we conclude that vacancies are an equilibrium feature in elemental crystals at temperatures above absolute zero. These are called *intrinsic* vacancies, because they are an inherent part of the crystal. If the only

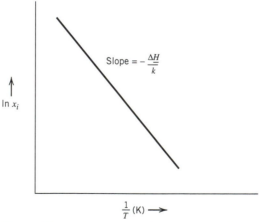

Figure 3.2 Fraction of imperfections (vacancies) as a function of inverse absolute temperature.

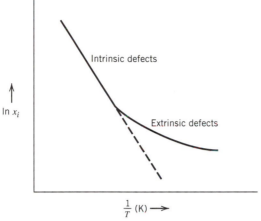

Figure 3.3 Fraction of imperfections (intrinsic and extrinsic) as a function of inverse absolute temperature.

vacancies present in a crystal are intrinsic vacancies, we will observe the linearity of ln x_i v $1/T$ shown in Figure 3.2. The concentration of vacancies may, however, be influenced by factors other than temperature.

The addition of impurities can, by distorting the lattice, create conditions near the impurity atoms that favor a vacancy concentration higher than the intrinsic concentration. The vacancies generated by such extraneous factors are called *extrinsic* vacancies. The presence of extrinsic vacancies usually manifests itself as a deviation from the linearity of ln x_i versus $1/T$ (Eq. 3.9) shown in Figure 3.2. Typically, this nonlinearity develops at lower temperatures (higher values of $1/T$), as shown in Figure 3.3. This chapter discusses several cases of extrinsic imperfections.

3.2   VACANCIES: EXPERIMENTAL VERIFICATION

In principle it should be possible to determine the values of $\Delta \underline{H}_v$ and $\Delta \underline{S}_v$ from a knowledge of the number of imperfections as a function of temperature. Unfortunately, it is difficult to measure the absolute number of imperfections. It is possible, however, to infer $\Delta \underline{H}_v$ and $\Delta \underline{S}_v$ by measuring the change in the number of imperfections with temperature. An especially novel approach to the problem was demonstrated successfully by Simmons and Balluffi (Refs. 1 and 2),[1] who asserted that

---

[1]Other methods have also been used to establish the temperature dependence of vacancy concentration. An early method used successfully (Refs. 3 and 4) involved quenching (rapid cooling) wires of the material being studied from high temperatures to a temperature at which the vacancies are immobile, usually below room temperature. The electrical resistivity of the wires was used as a measure of vacancy concentration. This method allows one to determine $\Delta \underline{H}_v$, but not $\Delta \underline{S}_v$. To determine both requires a measurement of the absolute number of vacancies at some temperature.

the volume of a crystal increases with increasing temperature for two reasons: the increased thermal vibration of the atoms and the creation of vacancies:

$$\Delta V = \Delta V_{Th} + \Delta V_v \tag{3.11}$$

The term ΔV_{Th} refers to the volume change because of increased thermal vibration of the atoms. The term ΔV_v refers to the volume change because of the presence of vacancies. It is equal to the number of vacancies multiplied by the volume change introduced by the presence of one vacancy, and is written as $\Delta V_v = n_v V_v$. To find the rate of change of ΔV_v with temperature at constant pressure, we substitute for n according to Eqs. 3.9 and 3.11, and differentiate with respect to temperature:

$$\left(\frac{\partial(\Delta V_v)}{\partial T}\right)_P = \left(\frac{\partial(n_v V_v)}{\partial T}\right)_P = N_A V_v A \exp\left[-\frac{\Delta \underline{H}_v}{kT}\right]\frac{\Delta \underline{H}_v}{kT^2}$$

Rearranging:

$$\frac{T^2}{V_v}\left[\frac{\partial \Delta V_v}{\partial T}\right]_P = \left[\frac{N \Delta \underline{H}_v A}{k}\right]\exp\left[-\frac{\Delta \underline{H}_v}{kT}\right] \tag{3.12}$$

Taking the natural logarithm of both sides of Eq. 3.12, we have

$$\ln\left[\frac{T^2}{V_v}\left(\frac{\partial \Delta V_v}{\partial T}\right)_P\right] = \ln\left[\frac{N \Delta \underline{H}_v A}{k}\right] - \frac{\Delta \underline{H}_v}{kT} \tag{3.13}$$

From Eq. 3.13, it is apparent that a graph of $\ln\left[\dfrac{T^2}{V_v}\left(\dfrac{\partial \Delta V_v}{\partial T}\right)_P\right]$ versus $1/T$ has a slope of $-\Delta \underline{H}_v/k$.

The value of the term $1/V\,(\partial V_v/\partial T)_P$ was determined by observing the difference between the change in external dimensions of a crystal $\Delta l/l$ and the change in lattice parameter $\Delta a/a$ as the temperature of the crystal is changed. In Eq. 3.14, the first term involving Δl is associated with the total volume change of the crystal. The term involving changes in lattice parameter, $\Delta a/a$, is related to the change of volume associated only with the increased vibrations of atoms. The difference between the two is the change in volume introduced by the formation of vacancies. Because the volume is a cubic function of a linear dimension of the crystal, the $\Delta V_v/V$ term is three times the fractional linear changes.

$$\frac{\Delta V_v}{V} = 3\left(\frac{\Delta l}{l} - \frac{\Delta a}{a}\right) \tag{3.14}$$

From these experiments Simmons and Balluffi inferred the enthalpies and entropies of formation of vacancies in aluminum, gold, and silver (Table 3.1). Based on

Table 3.1 Enthalpies and Entropies of Formation of Vacancies in Aluminum, Gold, and Silver

	$\Delta \underline{H}_v$ (eV)	$\Delta \underline{S}_v/k$	$\Delta n/n$ (at melting point)
Aluminum	0.75	2.2	9×10^{-4}
Gold	0.94	1.0	7.2×10^{-4}
Silver	1.09	1.5	2×10^{-4}

Source: Refs. 1 and 2.

their values, the calculated fraction of vacant lattice sites ($\Delta n/n$) at the melting point of these metals, also in Table 3.1, is shown to be almost one in a thousand. Let us use the values for silver to calculate the fraction of vacant sites at 700 K. The enthalpy and entropy values from Table 3.1 are $\Delta \underline{H}_v = 1.09$ EV and $\Delta \underline{S}_v/k = 1.5$.

$$\frac{n^*}{N} = \exp\left(\frac{\Delta \underline{S}_v}{k}\right) \exp\left(-\frac{\Delta \underline{H}_v}{kT}\right)$$

$$\frac{n^*}{N} = \exp(1.5) \exp\left(\frac{-1.09}{(8.617 \times 10^{-5})(700)}\right) = 64 \times 10^{-9}$$

The fraction of vacant lattice sites is thus about 64 parts per billion. The concentration of vacancies is the fraction of vacant sites multiplied by the number of lattice sites per unit volume, N_A/V. For silver, this calculation yields about 3.75×10^{15} vacancies per cubic centimeter at 700 K.

3.3   INTERACTIONS BETWEEN VACANCIES AND IMPURITIES

We have established that vacancies exist at equilibrium in pure, elemental crystals above absolute zero of temperature. The concentration of these intrinsic vacancies can be calculated using Eq. 3.8, if the energies and entropies of formation are known. The presence of dissolved impurity atoms in the crystal may influence the intrinsic vacancy concentration because the impurity atoms and the solvent atoms differ in size, causing vacancies to be attracted to the impurity atoms. The total concentration of vacancies would thus be modified.

To analyze this situation using statistical thermodynamics, consider that a vacancy can exist in two stable states, either bound to an impurity atom (extrinsic) or unbound (intrinsic). The total concentration of vacancies n_v is the sum of the two:

$$n_v = n_v^* + n_{I-v} \tag{3.15}$$

where n_v is total vacancy concentration (vacancies per unit volume), n_v^* is intrinsic vacancy concentration, and n_{I-v} is bound vacancy concentration.

The energy state of a vacancy depends on whether it is bound or unbound. Let us designate the energy of an unbound vacancy as E_0 and the energy of a bound vacancy as E_{I-v}. We may, for the purpose of this analysis, assign a value of zero to the unbound energy: $E_0 = 0$. The value of E_{I-v} will be inherently negative because it is a binding energy; that is, an energy input will be required to separate the vacancy from the impurity to create an unbound vacancy. Based on Eq. 2.7, the ratio of bound to unbound vacancies is

$$\frac{n_{I-v}}{n_v^*} = \frac{g_{I-v} \exp\left(-\dfrac{E_{I-v}}{kT}\right)}{Z_{pf}}$$

where Z_{pf} (the partition function) $= g^* + g_{Iv} \exp(-E_{I-v}/kT)$.

$$\frac{n_{I-v}}{n^*} = \frac{g_{I-v} \exp(-E_{I-v}/kT)}{g^* + g_{I-v} \exp(-E_{I-v}/kT)} \tag{3.16}$$

The value of the degeneracy, g_{I-v}, is the product of the impurity concentration, n_i, and Z, the coordination number in the crystal, because the vacancy may attach itself to an impurity at any of the nearest neighbor positions.[2] The degeneracy of the unbound vacancies is the number of lattice sites not nearest to impurity atoms.

$$g_{I-v} = Zn_i; \qquad g^* = N - n_i - Zn_i = N - (Z + 1)n_i$$

$$\frac{n_{I-v}}{n^*} = \frac{Zn_i \exp(-E_{I-v}/kT)}{N - (Z + 1)n_i + Zn_i \exp(-E_{I-v}/kT)}$$

Note that $N \gg (Z + 1)n_i$, and $Zn_i \exp(-E_{I-v}/kT) \ll N$ for small n_i:

$$n_{I-v} = n_v^* \frac{n_i}{N} Z \exp\left(-\frac{E_{I-v}}{kT}\right) \tag{3.17}$$

Substituting in Eq. 3.15, we write

$$\frac{n_v}{n_v^*} = 1 + \frac{n_i}{N} Z \exp\left(-\frac{E_{I-v}}{kT}\right) \tag{3.18}$$

To learn how important the extrinsic concentration can be, let us estimate some of the terms in Eq. 3.18. Let us take silver at 700 K once again as an example. Our calculation in Section 3.2 tells us that vacancies in silver are present at about 64

---

[2]The common notation for the coordination number is usually the letter Z, which we have used to signify the partition function. To avoid confusion, the notation Z_{pf} was used in this section for the partition function.

parts per billion at this temperature. It is difficult to reduce impurities in silver (n_i) to this level. A typical impurity fraction is 10^{-3}, or 0.1%. At this level, assuming a reasonable binding energy ($E_{I-v} = -0.1$ eV), and a coordination number of 12, the ratio of total vacancies to intrinsic vacancies[3] is:

$$\frac{n_v}{n^*} = 1 + 10^{-3}(12) \left(\exp \left[\frac{0.1}{(8.617 \times 10^{-5})(700)} \right] \right)$$

$$\frac{n_v}{n^*} = 1.06$$

At a 1% impurity level, this ratio rises to 1.63. This leads us to the conclusion that impurities can have an effect on vacancy concentration. Based on Eq. 3.18, this effect becomes even more important at lower temperatures.

3.4 INTERACTION BETWEEN IMPERFECTIONS AND IMPURITIES

In Section 3.3 we analyzed the effect of dissolved impurity atoms on the equilibrium concentration of point defects in a crystal, based on the formation of impurity–vacancy clusters. If there is a tendency for impurity atoms and crystal imperfections to form clusters, we should expect structural imperfections, such as dislocations, to attract and bind impurity atoms in a crystal. An interesting example of this binding is to be found in iron–carbon (steel) alloys. An accepted explanation for the "yield point" in steel is based on the binding between dislocations and solutes that occupy interstitial positions in the iron lattice, such as carbon atoms. Plastic deformation in a crystalline solid occurs when dislocations move under applied stress. In iron–carbon alloys, the stress required to start the movement of dislocations is higher than the stress required to keep them moving, giving rise to a distinct yield point (Ref. 5). The binding between dislocations and carbon atoms causes this phenomenon. The shear stress required to break the dislocation away from the relatively immobile carbon atoms is greater than the shear stress required to keep them moving. Based on reasoning similar to that used to derive Eq. 3.18, there should be fewer atoms bound to the imperfections as the temperature is increased. Thus, the yield point phenomenon should become less pronounced at higher temperatures. This is, indeed, what is observed in the iron–carbon system. At temperatures above 700°C the yield point essentially disappears.

To deal with the effect of binding of interstitial impurities to dislocations, let us adopt the same type of notation we used in Section 3.3. We can write:

$$\frac{n_{N\perp}}{n_N^*} = \frac{g_{N\perp} \exp(-E_{N\perp}/kT)}{g_N + g_{N\perp} \exp(-E_{N\perp}/kT)} \tag{3.19}$$

---

[3]The Boltzmann constant is 8.617×10^{-5} eV/K.

where $n_N^* = $ number of intrinsic nitrogen atoms in iron per cubic centimeter
   (intrinsic solubility)
$n_{N\perp} = $ number of nitrogen atoms bound to dislocations
$g_N = $ number of lattice positions available to dissolved nitrogen atoms
$g_{N\perp} = $ number of dislocation positions available to dissolved nitrogen
   atoms
$E_{N\perp} = $ interaction energy of dissolved nitrogen and dislocations
   relative to dissolved nitrogen in lattice (negative quantity)

Note that we have assigned an energy of zero to the dissolved nitrogen in a normal
position (interstitial). If we can assume that $g_N \gg g_{N\perp}$, Eq. 3.19 becomes

$$\frac{n_{N\perp}}{n_N^*} = \frac{g_{N\perp} \exp(-E_{N\perp}/kT)}{g_N} \tag{3.20}$$

The total number of nitrogen atoms dissolved n_T is the sum of n_N^* and $n_{N\perp}$.

$$n_T = n_N^* + n_{N\perp} = n_N^* \left(1 + \frac{g_{N\perp} \exp(-E_{N\perp}/kT)}{g_N}\right) \tag{3.21}$$

It is apparent from Eq. 3.21 that the measured solubility of nitrogen in iron is a
function of the dislocation concentration if the interaction energy $E_{N\perp}$ is not zero.
Let us estimate the magnitude of the effect. Interaction energies $E_{N\perp}$, are about -0.3
eV. If a sample is severely cold-worked (rolled), dislocation densities up to 5×10^{12}
cm/cm^3 can be reached. For iron, the number of interstitial positions per cubic centimeter is 8.4×10^{22}. The number of atoms per centimeter is the cube root of
8.4×10^{22}, or 4.3×10^7. Assuming, for simplicity, that we are dealing with edge
dislocations, the number of dislocation positions that can occupy a nitrogen atom is
the product of the dislocation density and 4.3×10^7, or 2.15×10^{20}. If we substitute
these values in Eq. 3.21 at 700 K (427°C), the enhancement of observed solubility
is about 37%. This effect should be observable, and, in fact, was observed by Darken
(Ref. 5).
Darken conducted his experiments by equilibrating samples of iron with a mixture
of ammonia and hydrogen at about 450°C. By controlling the ratio of ammonia to
hydrogen, he fixed the thermodynamic activity of nitrogen, because the equilibrium
constant K_a is determined by the temperature:

$$NH_3 = \tfrac{3}{2}H_2 + \underline{N} \tag{3.22}$$

$$K_a = a_{\underline{N}} \frac{P_{H_2}^{3/2}}{P_{NH_3}}$$

where underscored N indicates dissolved nitrogen.
The number of dislocations was varied by cold working the samples to different
levels. The greater the degree of cold work, the higher the dislocation concentration.

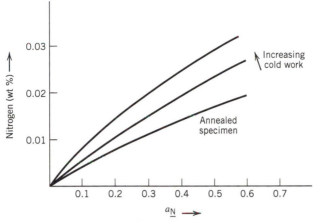

Figure 3.4 Nitrogen dissolved in iron as a function of activity of nitrogen and cold work (imperfections) in iron.

After equilibrating the iron samples with the gas of controlled nitrogen activity, Darken found higher nitrogen solubility in samples with greater cold work (i.e., the ones with higher dislocation densities). His observations are shown in Figure 3.4.

3.5 ELECTRONIC DEFECTS

Solid materials can, in broad terms, be divided into three classes with respect to their ability of conduct electrical charges: metals, semiconductors, and insulators. Metals have a high mobile carrier (electron) concentration, on the order of 10^{22} per cubic centimeter. At the other end of the scale, insulators have very few mobile charge carriers—on the order of one per cubic centimeter. The behavior of electrons in metals is discussed in texts on metal physics (see, e.g., Refs. 6 and 7). The conductivity of insulators is not discussed because, by the nature of the devices, this property cannot be varied significantly. Semiconductors are particularly interesting because their conductivity can be varied by changing the environment in which they exist. They are, thus, useful as sensors.

Intrinsic semiconductors are characterized by an energy gap (band gap) between electrons in their valence bands and allowed states in their conduction bands (Figure 3.5). To become active, in the sense of conduction, electrons must jump across the energy gap into the conduction band. When they do, they leave behind holes in the valence band. Both electrons and holes can participate in the conduction of electrical charge. Semiconductors can have a range of carrier concentrations, but for comparison with insulators and metals we can think of them as having on the order of 10^{10} mobile carriers per cubic centimeter. They are useful in electrical devices, such as transistors, or, in the case of ionic solids, as sensors, because the number of carriers changes with temperature or with the chemical nature of the atmosphere in equilibrium with the solid. In this section, we establish the relationship between charge

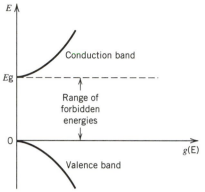

Figure 3.5 Plot of density of states $g(E)$ versus E for conduction electrons and holes in a semiconductor.

carriers in pure elemental crystals (electrons and holes) as a function of temperature and energy gap of the solid. A later section treats the effect of atmosphere.

If the two states for electrons, at the top of the valence band and at the bottom of the conduction band, had no limits as to their occupancy, calculating the probability of finding an electron at E_g relative to $E = 0$ (top of the valence band) would be a straightforward application of Boltzmann statistics (Eq. 2.7). The situation is more complicated, however, because the Pauli exclusion principle applies. Only one electron may occupy each quantum state in the crystal, taking all the quantum numbers into account, including the spins. We must therefore use the Fermi–Dirac statistics (Section 2.11). The number of electrons in the conduction band N_c is, based on Eq. 2.25:

$$N_c = \int_{E_g}^{\infty} g(E) \left[1 + \exp\left(\frac{E - \mu}{kT}\right) \right]^{-1} dE \tag{3.23}$$

where μ is the Fermi level in the crystal, E_F, and $g(E)$ is the density of states between energy levels E and E + dE.

The density of states for the electrons in the conduction band may be derived using the same technique we used to determine the density of states for an ideal gas, that is, from the "quantum particle in a box" model. Taking into account the two spin states that may exist at each energy level, the result for electrons is

$$g(E) = 4\pi \left(\frac{2m_e^*}{h^2}\right)^{3/2} (E - E_g)^{1/2}$$

$$g(E) = C_e(E - E_g)^{1/2} \tag{3.24}$$

where $C_e = 4\pi(2m_e^*/h^2)^{3/2}$ and m_e^* is the effective mass of an electron in the crystal.

For holes in the valence band, a similar approach yields:

$$g(E) = 4\pi \left(\frac{2m_h^*}{h^2}\right)^{3/2} (-E)^{1/2}$$

$$g(E) = C_h(-E)^{1/2}$$

(3.25)

where m_h^* is the effective mass of a ''hole'' in the valence band.

Because the Fermi level lies in the forbidden band, and its distance from the band edge is large compared with kT (which is equal to about 0.025 eV at room temperature), we may approximate the Fermi function as follows:

$$1 + \exp\left(\frac{E - E_F}{kT}\right) \approx \exp\left(\frac{E - E_F}{kT}\right)$$

(3.26)

because $(E - E_F) \gg kT$.

Substituting Eqs. 3.26 and 3.24 in 3.23 yields:

$$N_c = C_e \int_{E_g}^{\infty} (E - E_g)^{1/2} \exp\left[-\frac{(E - E_F)}{kT}\right] dE$$

(3.27)

If we let $x = (E - E_g)/kT$, Eq. 3.27 becomes

$$N_c = C_e(kT)^{3/2} \exp\left[-\frac{(E_g - E_F)}{kT}\right] \int_0^{\infty} x^{1/2}e^{-x}dx$$

For the definite integral, we write

$$\int_0^{\infty} x^{1/2}e^{-x}dx = \tfrac{1}{2}\sqrt{\pi}$$

hence:

$$N_c = 2\left(2\pi\,\frac{m_e^*kT}{h^2}\right)^{3/2} \exp\left[-\frac{E_g - E_F}{kT}\right]$$

(3.28)

By a similar process, the number of holes present in the valence band is,

$$N_h = 2\left(2\pi\,\frac{m_h^*kT}{h^2}\right)^{3/2} \exp\left(-\frac{E_F}{kT}\right)$$

(3.29)

If we are dealing with an intrinsic semiconductor—that is, one in which there is a hole left behind for every electron in the conduction band (no donor or acceptor impurities)—then,[4]

$$N_c = N_h \tag{3.30}$$

Substituting Eqs. 3.28 and 3.29 in Eq. 3.30:

$$E_F = \frac{E_g}{2} + \frac{3}{4} kT \ln \frac{m_h^*}{m_e^*} \tag{3.31}$$

Since kT is small, and the values of m_e^* and m_h^* are not greatly different, the Fermi level may be considered to be in the middle of the energy gap for intrinsic semiconductors.

If we think of the formation of conduction electrons and holes as a chemical reaction, we can write:

$$null = N_c + N_h \tag{3.32}$$

The product of N_c and N_h is a function of temperature, much as the equilibrium constant is, and has a value:

$$N_c N_h = 4 \left(\frac{2\pi k}{h^2} \right)^3 (m_e^* m_h^*)^{3/2} T^3 \exp \left(-\frac{E_g}{kT} \right) \tag{3.33}$$

It is interesting to note that the term for the Fermi level, E_F, drops out of Eq. 3.33, and the product of N_c and N_h depends only on temperature and the energy gap, E_g. In contrast to a simple, two-energy-level situation in which the Boltzmann distribution (Eq. 2.7) would apply, the product of the number of electrons and holes is in this case a function of E_g/kT, but also a function of T^3.

3.6   DEFECTS IN IONIC COMPOUNDS

The preceding sections dealt with defects in elemental crystals, where the atoms were all of the same species. Ionic compounds have a minimum of two elements, an electropositive element, designated by M, and an electronegative element, designated by X. Vacancies may occur on either the electropositive or the electronegative lattice. Additionally, both the electropositive and the electronegative elements may take interstitial positions in the lattice. Each type of defect is discussed in the sections that follow.

---

[4]In many publications, N_c, the number of electrons in the conduction band, is written simply as "n." The number of holes N_h is written as "p," referring to a positively charged conduction species.

Table 3.2 Summary of Kroger–Vink Notation

Symbol	Definition
M	Atom of electropositive element
X	Atom of electronegative element
M_M	M atom on M site (sometimes denoted as M_M^x)
N_M	N atom on M site
V_M	Vacancy on M site
M_i	M atom on interstitial site
$M_i^{\bullet}$	Positively charged M ion on interstitial site (singly ionized)
$M_i^{\bullet\bullet}$	Positively charged M ion on interstitial site (doubly ionized)
V_i'	Negatively charged X ion on interstitial site (singly ionized)
$V_X^{\bullet}$	Positively charged vacancy (relative to perfect lattice) on X site
V_M'	Negatively charged vacancy (relative to perfect lattice) on M site

Source: Ref. 8.

For the purposes of our discussion, we will adopt the Kroger–Vink notation (Ref. 8) to represent these imperfections (Table 3.2). An easy way (Ref. 9) to understand the notation is to consider that each point defect is represented by a three-part symbol, X_Y^Z, where

X represents *what* is on the site (V for a vacancy and an elemental notation if the site is occupied by an element)

Y represents the *type* of site occupied by X (i for an interstitial site, or, for a lattice site, the symbol for the element usually occupying that lattice site)

Z represents the *charge* relative to the normal ionic charge on the site Y [positive charges are represented by dots ($\bullet$ or $\bullet\bullet$), and negative charges by primes (' or "); a lowercase letter x is sometimes used to denote a neutral atom].

In silver chloride, AgCl, an interstitial silver ion is represented as $Ag_i^{\bullet}$. A vacancy on the silver lattice is represented as V_{Ag}'.

Equations describing defect formation or annihilation must adhere to the following principles:

1. Conservation of mass
2. Conservation of charge (electroneutrality)
3. A fixed ratio of M and X sites according to the compound being studied

3.7 FRENKEL DEFECTS

One type of defect in an ionic solid is formed when an ion normally found on one of the lattice positions lies in an interstitial position in the structure, and a lattice position corresponding to that element is vacant. This is called a *Frenkel defect*. The formation of a Frenkel defect by an electropositive element (M) may be expressed as a chemical equation:

$$MX = M_{(1-\delta)}X + \delta V_M + \delta M_i \qquad (3.34)$$

In the Kroger–Vink notation, which we will use, this is written as follows:

$$M_M = M_i^\cdot + V_M' \qquad (3.35)$$

We have assumed that the interstitial M atom is ionized with a charge of $+1$. The vacancy has a charge of -1 relative to the perfect lattice. If we think of the change expressed in Eq. 3.35 as a chemical reaction, we can write an equilibrium constant for it:

$$K_1 = \frac{a_{M_i^\cdot} a_{V_M'}}{a_{M_M}} \qquad (3.36)$$

The activity of the material M (a_{M_M}) is very close to unity because only small quantities of defects are formed. Using the infinitely dilute solution as the standard state for the imperfections makes their activities equal to their concentrations[5]:

$$K_1 = [M_i^\cdot][V_M'] \qquad (3.37)$$

To describe the concentration of these imperfections as a function of temperature, we follow the same procedure we used for elemental crystals, with the complication that we must account for the configurational entropy ($\Delta \underline{S}_c$) of both the vacancies and the interstitials. There is uncertainty concerning the position of the vacancies and also uncertainty related to the position of the interstitials. If we assume that the locations of the vacancies and the interstitials are random and independent of one another, we can write

$$\Delta \underline{S}_c = k \ln \left[\frac{(N + n_v)!}{N! n_v!} \frac{(N + n_i)!}{N! n_i!} \right] \qquad (3.38)$$

where n_v = number of vacancies
n_i = number of interstitials

Proceeding as in Section 3.1, we obtain:

$$\left(\frac{n_v}{N + n_v} \right) \cdot \left(\frac{n_i}{N + n_i} \right) = \exp \left(\frac{\Delta \underline{S}_f}{k} \right) \exp \left(- \frac{\Delta \underline{H}_f}{kT} \right) \qquad (3.39)$$

---

[5]Square brackets [ ] denote the concentration of the species in the brackets.

but

$$n_v \ll N; \quad n_i \ll N; \quad [V'_M] = \frac{n_v}{V}; \quad [M_i^\bullet] = \frac{n_i}{V}$$

For one mole of crystal:

$$[V'_M][M_i^\bullet] = \frac{N_A^2}{V^2} \exp\left(\frac{\Delta S_f}{k}\right) \exp\left(-\frac{\Delta H_f}{kT}\right) \tag{3.40}$$

The Frenkel defect concentration, [F] is equal to the concentration of either the interstitials $[X_i]$ or the vacancies [V], if the two are equal, or to the square root of their product:

$$[F] = ([V'_M][M_i^\bullet])^{1/2} = \frac{N_A}{V} \exp\left(\frac{\Delta S_f}{2k}\right) \exp\left(-\frac{\Delta H_f}{2kT}\right) \tag{3.41}$$

This term, [F], is called the intrinsic Frenkel defect concentration and is related to the entropy and enthalpy of formation of the defect.

The preceding illustration was based on Frenkel defect formation by the electropositive (M) elements. The same can also occur for the electronegative (X) elements.

3.8   SCHOTTKY–WAGNER DEFECTS

A defect in an ionic crystal may be created by having an electrical-charge-equivalent number of vacancies created on the electropositive and electronegative lattices. If the electropositive and electronegative elements have the same valence, there will be the same number of vacancies on each lattice. If the valence of the two elements is different, the number of vacancies is inversely proportional to their valence states. In the case of similar valence this can be expressed as a chemical equation as follows (the term ''null'' represents the defect-free lattice):

$$\text{null} = V_X^\bullet + V'_M \tag{3.42}$$

In terms of an equilibrium constant, this can be written as follows:

$$K_2 = [V_X^\bullet][V'_M] \tag{3.43}$$

Note that the term $[M'_M]$ appears both in Eqs. 3.37 and 3.43. Both Frenkel and Schottky–Wagner defects may be present in the same crystal. Chemical equilibrium requires that both equilibria be satisfied in addition to the condition of stoichiometry, which in this case corresponds to electroneutrality among the charged species. Because of these conditions, we may write:

$$[V_M'] = [M_i^{\cdot}] + [V_X^{\cdot}] \tag{3.44}$$

$$[M_i^{\cdot}] = \frac{K_1}{[V_M']} \; ; \; [V_X^{\cdot}] = \frac{K_2}{[V_M']}$$

$$[V_M']^2 = K_1 + K_2 \tag{3.45}$$

$$[V_M'] = (K_1 + K_2)^{1/2}$$

Thus the concentration of vacancies on the electropositive lattice depends on the presence of both Frenkel and Schottky–Wagner defects.

3.9   INTERACTIONS AMONG DEFECTS

In Section 3.7 we demonstrated that the concentrations of Frenkel and Schottky–Wagner defects are interdependent. This interdependence of defect concentrations also holds for a wide variety of other defects. In this section we discuss a useful way of representing these relationships in diagrammatic form. In the literature of the field, the diagrams are called either Brouwer diagrams or Kroger–Vink diagrams (Refs. 8 and 10).

Take as an example the case of an ionic solid, represented as MX, which is exposed to the vapor of the metal M at various pressures. Let us assume that the metal vapor interacts with the MX crystal to form interstitial ions and electrons, which we will label n':

$$M(g) = M_i^{\cdot} + n'$$

The equilibrium among the species in the reaction is represented by

$$K_F = \frac{[n'][M_i^{\cdot}]}{P_M} \tag{3.46}$$

We note that there are also charged vacancies on the M sublattice, and we must accommodate to the Frenkel defect equilibrium:

$$K_F' = [M_i^{\cdot}][V_M'] \tag{3.47}$$

In addition, there is an equilibrium between free electrons (n') and holes (p'):

$$K_i = [n'][p^{\cdot}] \tag{3.48}$$

Electroneutrality of the crystal imposes the following condition:

$$[n'] + [V_M'] = [p^{\cdot}] + [M_i^{\cdot}] \tag{3.49}$$

If we take the natural logarithms of the terms in Eqs. 3.46, 3.47, and 3.48, we obtain:

$$\ln K_F = \ln[M_i^{\cdot}] + \ln[n'] - \ln P_M \tag{3.50}$$

$$\ln K_F' = \ln[M_i^{\cdot}] + \ln[V_M'] \tag{3.51}$$

$$\ln K_i = \ln[n'] + \ln[p^{\cdot}] \tag{3.52}$$

The construction of a Brouwer (or Kroger–Vink) diagram proceeds by noting that certain approximations may be made in different ranges of metal vapor pressure values. For example, in a low metal pressure range we may assume that the concentrations of interstitial cations and holes predominate. Based on the condition of electroneutrality (Eq. 3.49) the concentrations of the two are equal:

$$[p^{\cdot}] = [V_M'] \tag{3.53}$$

As the pressure in the metal vapor increases, we enter a region of stoichiometry in which the concentrations of interstitial cations and negatively charged vacancies are equal (Frenkel defect):

$$[V_M'] = [M_i^{\cdot}] \tag{3.54}$$

Finally, as the pressure increases, we enter a region where the concentration of cation interstitials is equal to the concentration of free electrons:

$$[n'] = [M_i^{\cdot}] \tag{3.55}$$

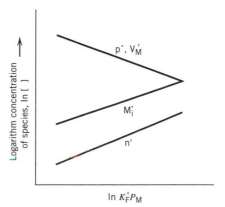

Figure 3.6 Concentration of charged species in MX exposed to M(g) as a function of metal vapor pressure.

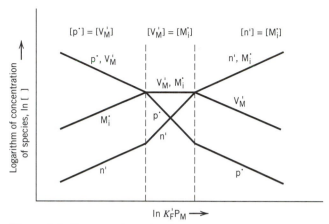

Figure 3.7 Concentration of various species in MX exposed to M(g) as a function of metal vapor pressure.

Treating the case represented by Eq. 3.53, we can write Eq. 3.50 as follows:

$$\ln P_M = \ln[M_i^{\cdot}] + \ln[n'] - \ln K_F \qquad (3.50')$$

Combining Eqs. 3.51 and 3.52 yields:

$$\ln \left(\frac{K_F'}{K_i}\right) = \ln[M_i^{\cdot}] - \ln[n'] \qquad (3.56)$$

Combining the results of Eqs. 3.50' and 3.56 yields:

$$\ln P_M = 2\ln[M_i^{\cdot}] - \ln \left(\frac{K_F' K_F}{K_i}\right) \qquad (3.57)$$

These relationships are shown schematically in Figure 3.6. Note that on the right-hand side of Figure 3.6 the concentration of cation interstitials becomes equal to the concentration of cation vacancies, and we enter the pressure region dominated by Frenkel defects. A further increase in gaseous metal vapor pressure moves us into the region dominated by free electrons and cation interstitials. The concentrations of the various species involved are shown schematically in Figure 3.7.

3.10   INTRINSIC AND EXTRINSIC DEFECTS IN IONIC CRYSTALS

Vacancies in ionic crystals exist at equilibrium at temperatures above absolute zero, as demonstrated in the preceding sections. These thermally induced vacancies in pure M—X crystals are called *intrinsic* vacancies. Vacancies may also be induced by impurity ions whose valence differs from the valence of the ions in the host crystal. These are called *extrinsic* vacancies.

Consider the addition of cadmium chloride ($CdCl_2$) to a sodium chloride (NaCl) crystal. From the chemical equations describing the compounds, it is apparent that the valence of the cadmium ion in cadmium chloride is +2. The sodium ion in sodium chloride has a valence of +1. *One* $CdCl_2$ molecule occupies *two* anion (negative ion) sites, and *one* cation site. Thus, in a dilute solution of cadmium chloride in sodium chloride, there must be one sodium ion vacancy on the M lattice for every cadmium ion added, because electrical neutrality is required (Figure 3.8). This condition of electroneutrality can be expressed as follows:

$$[V'_{Na}] = [Cd^{\cdot}_{Na}] + [V^{\cdot}_{Cl}] \qquad (3.58)$$

The concentration of vacancies on the sodium ion lattice is equal to the concentration of vacancies in the chlorine lattice plus the concentration of cadmium ions.

From Eq. 3.58 it is clear that if the concentration of cadmium ions is very much greater than the vacancies of chloride ions ($[Cd^{\cdot}_{Na}] \gg [V^{\cdot}_{Cl}]$), the vacancy in sodium ion concentrations simply equals the concentration of cadmium ions. In this case, the concentration of sodium ion vacancies is *extrinsic;* that is, it is not controlled by the inherent properties of the sodium chloride. It depends on another condition (viz., the cadmium ion concentration) and is not related to the *intrinsic* number of vacancies produced thermally. If, however, the cadmium ion concentration is very small ($[V^{\cdot}_{Cl}] \gg [Cd^{\cdot}_{Na}]$), the concentration of sodium ion vacancies is equal to the concentration of chloride ion vacancies, and the concentrations are intrinsic—that is, only thermally induced.

These relationships can be illustrated using the equilibrium relationship for Schottky defect on the sodium chloride lattice:

$$K_S = [V'_{Na}][V^{\cdot}_{Cl}] \qquad (3.59)$$

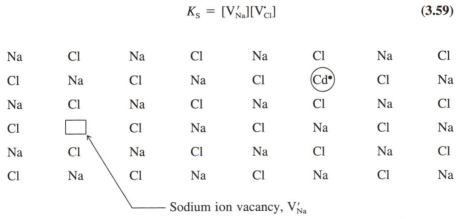

Figure **3.8** Effect of $CdCl_2$ addition (presence of $Cd^{\cdot}_{Na}$) on NaCl crystal.

Combining Eqs. 3.59 and 3.58 yields:

$$[V'_{Na}] = [Cd^{\cdot}_{Na}] + \frac{K_S}{[V'_{Na}]}$$

or

$$[V'_{Na}]^2 - [Cd^{\cdot}_{Na}][V'_{Na}] - K_S = 0$$

The solution to this equation is:

$$[V'_{Na}] = \frac{[Cd^{\cdot}_{Na}] + ([Cd^{\cdot}_{Na}]^2 + 4K_S)^{1/2}}{2} \tag{3.60}$$

A sodium chloride crystal with a given concentration of cadmium chloride can display both intrinsic and extrinsic behavior depending on its temperature. This can be understood by considering Eq. 3.60. At higher temperatures, the value of the equilibrium constant K_S can be much greater than the value of $[Cd^{\cdot Na}]^2$. If that is true, then $[V'_{Na}] = K_S^{1/2}$ and the thermally generated imperfections will be greater than the cadmium ion concentration. The crystal will show intrinsic behavior. As the temperature of the crystal drops, the value of K_S drops; that is, the number of thermally induced vacancies drops. When K_S is much smaller than $[Cd^{\cdot}_{Na}]^2$, we will see $[V'_{Na}] = [Cd^{\cdot}_{Na}]$ and extrinsic behavior. This change in behavior with temperature

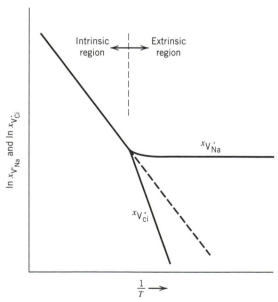

Figure 3.9 Intrinsic and extrinsic vacancy regions.

is illustrated in Figure 3.9, which plots the natural log of a sodium ion vacancy against the inverse absolute temperature.

In the intrinsic region, the vacancies in sodium ions are equal to the vacancies in chloride ions. In the extrinsic region, the equilibrium expression (from Eq. 3.43) must still be observed; thus the vacancies in chloride ions will be less than the quantity calculated for intrinsic chloride ion concentration as illustrated in Figure 3.9.

3.11 EXPERIMENTAL DETERMINATION OF DEFECT TYPE

The different defect types in ionic solids discussed so far should manifest themselves in physically observable ways. In fact, if the ionic solids are to be used as sensors, we want to observe a variation of some measurable quantity as a function of temperature, or of the environment in which the solid exists. This section discusses two ways of inferring defect type in ionic crystals, one based on the measurement of lattice parameters and densities, and the other based on the measurement of electrical conductivity.

As an example of the use of density and lattice parameter measurements, consider the case of the addition of calcia (CaO) to zirconia (ZrO_2). If 15 mol % calcia is incorporated in the zirconia lattice, and assuming that the valences of the ions are unchanged at Ca^{2+}, Zr^{4+}, and O^{2-}, electroneutrality among the three ionic species present yields a composition of $Zr_{0.85}Ca_{0.15}O_{1.85}$. There are two ways of forming this crystal. Either vacancies form on the oxygen lattice, or the calcium and/or zirconium ions enter interstitial positions. It is known from X-ray diffraction studies that the material crystallizes in a fluorite structure with a lattice parameter of 5.131 Å. Let us calculate the expected densities of the crystal for the two cases. If oxygen ion vacancies were to form, the unit cell weight would be $452.60/N_A$, which is the sum of:

$$\left.\begin{array}{ll} \text{Ca:} & \dfrac{0.15 \times 4 \times 40.08}{N_A} \\[2ex] \text{Zr:} & \dfrac{0.85 \times 4 \times 91.22}{N_A} \\[2ex] \text{O:} & \dfrac{1.85 \times 4 \times 16}{N_A} \end{array}\right\} = \dfrac{452.60}{N_A}$$

The volume of the unit cell is the cube of the lattice parameter, or 135.08 Å³, or 135.08×10^{-24} cm³.

The density would be:

$$\rho = \frac{452.60}{(6.022 \times 10^{23})(135.08 \times 10^{-24})} = 5.57 \text{ g/cm}^3$$

If interstitial calcium and zirconium ions formed, then the weight of the unit cell would be $489.29/N_A$ which is the sum of:

$$
\left.
\begin{aligned}
&\text{O:} && \frac{8 \times 16}{N_A} \\[2mm]
&\text{Ca:} && \frac{4 \times 0.15 \times (2/1.85) \times 40.08}{N_A} \\[2mm]
&\text{Zr:} && \frac{4 \times 0.85 \times (2/1.85) \times 91.22}{N_A}
\end{aligned}
\right\} = \frac{489.29}{N_A}
$$

The density would be:

$$
\rho = \frac{489.29}{(6.022 \times 10^{23})(135.08 \times 10^{-24})} = 6.01 \text{ g/cm}^3
$$

The two values 5.57 and 6.01 g/cm^3 can be easily distinguished by density measurements. In this case, oxide ion vacancies are found to be present in CaO-ZrO$_2$ solutions equilibrated at 1600°C.[6] This is an especially important conclusion, because "doped" crystals of zirconia are useful as oxide ion conductors in high temperature fuel cells and as oxygen pressure sensors when used in electrochemical cells (Vol. I, section 6.8).

Electrical conductivity measurements also can be used to determine the types of defect present in ionic crystals. Consider the case of zinc oxide (ZnO). If zinc oxide is heated at low oxygen pressures, it will become conductive. By Hall effect measurements (Refs. 6 and 7), it can be shown that the conducting species is negatively charged, and by the magnitude of the conductivity it is assumed that it is electrons (n') that are the charge carriers. Let us hypothesize that the defect structure is of the Frenkel type, with zinc ions occupying the interstitial position. The question is whether an interstitial zinc ion in the sample is singly or doubly charged. If it is singly charged, then the chemical reaction is:

$$
\text{ZnO} = \tfrac{1}{2}\text{O}_2 + \text{Zn}_i^{\cdot} + \text{n}' \tag{3.61}
$$

The equilibrium constant for the reaction is:

$$
\text{K} = P_{\text{O}_2}^{1/2}[\text{Zn}_i^{\cdot}][\text{n}'] \tag{3.62}
$$

Noting that the concentration of zinc ions must equal the concentration of electrons,

$$
[\text{Zn}_i^{\cdot}] = [\text{n}'] \tag{3.63}
$$

---

[6]For samples equilibrated at 1800°C there seems to be some ambiguity in the type of defect present (Ref. 11).

then the concentration of electrons, and the conductivity, will be proportional to the negative one-fourth power of the oxygen pressure, as follows:

$$[n'] = [Zn_i^{\cdot}] = K^{1/2}P_{O_2}^{-1/4} \tag{3.64}$$

If the interstitial zinc ion is doubly charged, concentration of electrons will be proportional to the negative one-sixth power of the oxygen pressure.

$$ZnO = \tfrac{1}{2}O_2 + Zn_i^{\cdot\cdot} + 2n' \tag{3.65}$$

$$[Zn_i^{\cdot\cdot}] = \tfrac{1}{2}[n']$$
$$\tag{3.66}$$
$$[n'] = \tfrac{1}{4}KP_{O_2}^{-1/6}$$

The conductivity of zinc oxide crystals is found to be proportional to the negative one-fourth power of the oxygen pressure, hence singly ionized interstitial zinc ions exist in ZnO.

3.12   NON STOICHIOMETRY

Strongly ionic compounds, such as those containing halides, generally display a fixed ratio of constituents. Sodium chloride, for example, shows a one-to-one ratio of cations to anions, cadmium chloride a two-to-one ratio. Many of the compounds studied in elementary chemistry are of this type. We have demonstrated that through exposure to different atmospheres, these ratios may be changed somewhat. In fact, there are compounds in which the ratios vary greatly from the expected stoichiometric ratios. A classic example is provided by one of the iron oxides, wustite.

The compound FeO (wustite) does not exist with a one-to-one cation-to-anion ratio and actually has a range of compositions (Figure 3.10). The iron-rich limit of the compound has a composition closer to $Fe_{0.95}O$. At 1200°C the oxygen content of the wustite may be varied by equilibrating samples in atmospheres with effective oxygen pressures between 10^{-9} and 10^{-12} atm. The departure from stoichiometry may be due either to the formation of oxygen ion interstitials or to cation vacancies. The data in Table 3.3 indicate that the nonstoichiometry is due to cation vacancies because the density of the oxide decreases as the ratio of iron to oxygen decreases. To maintain electroneutrality of the compound, two of the Fe^{2+} ions must be transformed into Fe^{3+} ions for each of the vacancies formed. In chemical terms, this may be considered to be a solution of Fe_2O_3 in FeO. Similar nonstoichiometric behavior is observed in many systems, such as Ni–O, Co–O, and Cu–O. Not all of the nonstoichiometry is accounted for by cation vacancies. In the Zr–O and Ti–O systems, anion vacancies form. As demonstrated in Section 3.10, interstitial cations account for nonstoichiometry in the Zn–O system. Thus we can conclude that the principle

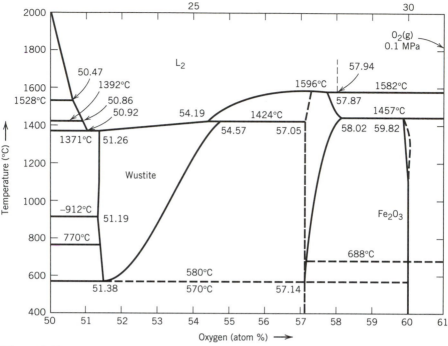

Figure 3.10 Phase diagram for Fe–O from 50 to 60 atom % O; condensed system, 0.1 MPa.

Table 3.3 Lattice Parameter and Density of FeO

Composition	Lattice Parameters	Density (g/cm³)
$Fe_{0.91}O$	4.282	5.613
$Fe_{0.92}O$	4.285	5.624
$Fe_{0.93}O$	4.292	5.658
$Fe_{0.945}O$	4.301	5.728

Source: E. R. Jette and F. Foote, *J. Chem. Phys.,* 1, 29 (1933), quoted in Ref 12.

of fixed ratios of atoms in molecules that simplified many of the considerations in elementary chemistry does not apply to many solid compounds.

REFERENCES

1. Simmons, R. O., and Balluffi, R. W., *Phys. Rev., 117,* 52 (1960).
2. Simmons, R. O., and Balluffi, R. W., *Phys. Rev., 119,* 600 (1960).
3. Desonbo, W., and Turnbull, D., *Acta Met., 7,* 83 (1959).
4. Bauerle, J. E., and Koehler, J. S., *Phys. Rev., 107,* 1493 (1957).

5. Darken, L. S., *National Physical Laboratory Symposium No. 9,* Paper 4G, Her Majesty's Stationery Office, London, 1959.
6. Solymar, L., and Walsh, D., *Lectures on the Electrical Properties of Materials,* Oxford University Press, New York, 1988.
7. Omar, M. A., *Elementary Solid State Physics,* Addison-Wesley, Reading, MA, 1975.
8. Kroger, F. A., and Vink, V. J., *Solid State Phys., 3,* 307–435 (1956).
9. Allen, S. M., and Thomas, E. L., class notes for course 313, Massachusetts Institute of Technology, 1994.
10. Brouwer, G., *Philips Res. Rep., 9,* 366 (1954).
11. Diness, A., and Roy, R., *Solid State Commun., 3,* 123 (1965).
12. Swalin, Richard A., *Thermodynamics of Solids,* second edition, Wiley, New York, 1972.

PROBLEMS

3.1 When CaO is dissolved in ZrO_2, oxide ion vacancies are created on the zirconia lattice. Calculate the fraction of oxide ion sites that will be vacant when one mole percent of calcium oxide is added to (i.e., dissolved in) the zirconia. Neglect the intrinsic oxide ion vacancies.

   What is the oxide ion vacancy concentration stated at vacancies per cubic centimeter? The density of zirconia is about 5.5 g/cm^3.

3.2 Silver bromide (AgBr) exhibits Frenkel imperfections. The formation enthalpy is about 0.60 eV per pair. Assume that the formation entropy is zero.

   (a) What fraction of the Ag^+ ions are in interstitial sites just below the melting point of silver bromide?

   (b) The Ag^+ ions are extremely mobile, permitting the establishment of equilibrium even at relatively low temperatures. Assume that a one cubic centimeter sample of an AgBr crystal is cooled slowly enough to maintain equilibrium. At what temperature will the crystal contain just one Frenkel pair?

For AgBr:

Density = 6.47 g/cm^3
Molecular weight = 187.8 g/mol
Melting point = 700 K.

3.3 The formation enthalpy of a Schottky defect in NaCl is about 1.87 eV.

   (a) Plot the fraction of Na^+ and Cl^- vacancies as a function of the reciprocal temperature from the melting point, 1074 K, to 300 K. Assume that the formation entropy of the vacancies is zero.

   (b) Repeat part a, assuming that 10^{-6} mole fraction of $CaCl_2$ is added to the pure NaCl. Why are the plots in parts a and b different?

3.4 Density measurements are often used to determine the type of defects created when impurities are dissolved in ionic crystals. If one assumes that the dissolution of aluminum oxide (Al_2O_3) in pure magnesium oxide (MgO) creates vacancies on the positive ion lattice, calculate the fractional change in the density of pure MgO when one weight percent of Al_2O_3 is added to it.

Assume that the crystal structure and the lattice parameter of MgO do not change with the addition of the alumina.

3.5 The energy of formation for vacancies in gold is about 0.9 eV. Assuming that there is no change in vibrational entropy associated with the formation of a vacancy, calculate the intrinsic vacancy concentration in gold at 500 and 1000 K.

Silver, when dissolved in gold, exhibits a binding energy of about 0.10 eV with vacancies. Calculate the total vacancy concentration in gold at 500 and 1000 K. Assume that there is no change in vibrational entropy involved in the formation of the silver–vacancy complex. The silver concentration is 0.1 mol percent.

The density of gold is about 19.3 g/cm³.

3.6 Magnesium oxide (MgO) is known to form Schottky defects, creating vacancies on both the cation (Mg) and anion (O) lattices. The energy of formation of a Schottky defect in MgO is estimated to be about 6 eV. Assume that we can consider the entropy of formation of the defect to be zero [the preexponential term, $\exp(\Delta S/2R) = 1$].

(a) Estimate the fraction of Mg sites and O sites vacant at equilibrium at 1800 K.

(b) What is the concentration of Mg vacancies (vacancies per cubic centimeter) at equilibrium at 1800 K?

(c) If a sample of MgO with 0.1 mol % of zirconia (ZrO_2) is prepared, what will be the fraction of Mg sites unoccupied at equilibrium at 1800 K? What will be the fraction of oxygen sites unoccupied? The substitution of zirconium ions (Zr^{+4}) in the lattice results in cation vacancies.

DATA

Density of MgO = about 3.58 g/cm³
Atomic weight of Mg = 24.3 g/mol
Atomic weight of oxygen = 16 g/mol

3.7 A sample of iron oxide (wustite) has a composition $Fe_{0.93}O$. Its lattice parameter is found to be 4.301 Å.

(a) Calculate the density of the sample (g/cm³), assuming that the nonstoichiometry of the compound is accounted for by vacancies on the Fe lattice.

(b) Calculate the density of the sample (g/cm³), assuming that the nonstoichiometry of the compound is accounted for by oxygen interstitials.

DATA

Atomic weights
Fe = 55.85 g/mol
O = 16 g/mol

FeO has a "rock salt" crystal structure (fcc) with 4 cations and 4 anions per unit cell in a perfect (defect-free) crystal.

3.8 Given the information below, you are asked to estimate the number of neutral (noncharged) vacancies in pure, metallic silicon at 1200 K.

(a) What fraction of the silicon sites are vacant at 1200 K (at equilibrium)?

(b) What is the concentration of these vacancies (per cubic centimeter)?

DATA

ΔH_f = 2.4 eV per vacancy
exp($+\Delta \underline{S}_f/k$) = 3.0 for the vacancy
Atomic weight of silicon = 28 g/cm³
Density of silicon = 2.42 g/cm³

3.9 The accompanying graph shows the self-diffusion coefficient for Na^+ in an NaCl crystal doped with $CdCl_2$. NaCl is believed to form Schottky defects. Using the data in the graph, calculate the following.

(a) The activation energy for self-diffusion of Na^+ in NaCl.

(b) The enthalpy of formation of Schottky defects in NaCl (in eV).

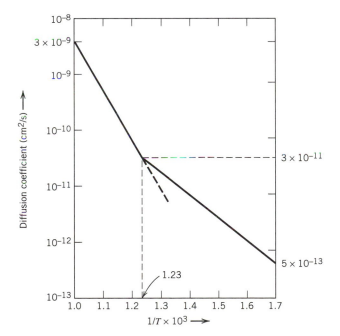

Chapter 4

Surfaces and Interfaces

Everything has to end somewhere. For solids, or liquids, that "somewhere" is a surface, an interface between phases. For liquids, the interface is between the liquid and a vapor phase, or between two immiscible liquids. In solids, the interface can be between the solid and a vapor or liquid phase, between two chemically different solids, or between crystals (grains) of the same chemical composition but differing in crystallographic orientation. The properties of these interfaces differ from bulk properties, and they influence the behavior of materials in many ways. This chapter explores some of the effects attributable to surfaces. A later chapter (Chapter 6) will discuss the influence of surfaces on the process of nucleation.

An atom at a free surface of a solid has greater energy than an atom in the interior of a crystal because it is less tightly bound. In a close-packed crystal, for example, an atom *inside* the crystal is bound to its 12 neighbors. An atom on a free *surface* is missing approximately three of these neighbors, hence it is bound to only nine atoms. The surface atoms, being less bound to others, exist at higher energy levels than atoms in the bulk of the crystal. In general terms, the sum of all the excess energies of the surface atoms (excess with respect to atoms inside the crystal) is the surface energy.

Surface energy can be defined in terms of energy, enthalpy, Helmholtz free energy, or Gibbs free energy, depending on the physical constraints placed on the

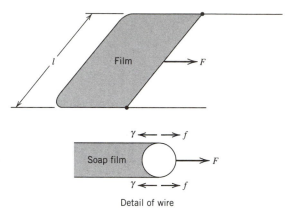

Detail of wire

Figure 4.1 Stretching a film—increasing area.

definition.[1] The most useful to us will be the one based on Gibbs free energy. Applying the principles in Chapter 1 (Eqs. 1.21, 1.22, and 1.28), the surface energy γ in
terms of the Gibbs free energy is:

$$dG = -S\,dT + V\,dP + \gamma\,dA \qquad\qquad \textbf{(4.1)}$$

$$\gamma = \left(\frac{\partial G}{\partial A}\right)_{T,P}$$

There are several commonly observed manifestations of this surface energy. As
a liquid droplet tries to minimize its free energy, it assumes a spherical shape—that
is, the shape that minimizes its surface area. Small droplets tend to agglomerate into
larger droplets, minimizing their combined surface-to-volume ratios. Small particles
of metals or ceramics, when pressed into a shape and heated at high temperatures,
will "sinter" into a solid mass. The driving force for this sintering process is the
reduction of surface energy.

4.1   SURFACE ENERGY–SURFACE TENSION

Imagine a film of liquid stretched on a frame as shown in Figure 4.1. The loop of
wire on the frame can be pulled to the right by a force, F, thereby creating new

---

[1]The surface energy in terms of other thermodynamic functions is:

$$\gamma = \left(\frac{\partial U}{\partial A}\right)_{S,V} = \left(\frac{\partial H}{\partial A}\right)_{S,P} = \left(\frac{\partial F}{\partial A}\right)_{T,V} = \left(\frac{\partial G}{\partial A}\right)_{T,P}$$

surface (two surfaces, the top and bottom of the film). The balance of forces and the work done can be expressed as follows:

$$2l\gamma\, dx = F\, dx$$

$$\gamma = \frac{F}{2l}$$

The units of surface tension, γ, are dynes per centimeter, or newtons per meter in the SI system.

By multiplying the numerator and denominator of γ by a length term (centimeters), the dimensions become dyne-centimeters per square centimeter, or ergs per square centimeter (joules per square meter in the SI system). Thus γ can be thought of as *surface tension* or *surface energy*. For liquids, the two are equal. For solids they are not.

The surface energies of solids and liquids differ in many ways. The surface energy of a liquid is isotropic; that is, it is independent of direction. The surface energy of a solid is a function of the crystallographic plane that is exposed. Furthermore, the surface tension of a solid varies with direction in the exposed plane. In a liquid, the unit surface energy does not change as the surface is stretched. As the surface of a liquid is increased, atoms of the liquid come to the surface from the bulk, maintaining the surface characteristics. In the case of solids, this does not occur at normal temperatures, where atomic mobility is limited. Hence, the nature of a solid surface changes as the material is deformed.

4.2 APPROXIMATE CALCULATION OF SOLID SURFACE ENERGY

The magnitude of solid surface energy can be approximated by a simple calculation. If we make the simplifying assumption that the binding energy of an atom to a solid is the result of discrete bonds to its nearest neighbors, then the energy of one bond, ε, can be written as follows:

$$\varepsilon = \frac{\Delta \underline{H}_s}{0.5ZN_A} \tag{4.2}$$

where $\Delta \underline{H}_s$ is the molar enthalpy of sublimation (breaking all the bonds), Z is the coordination number, and N_A is Avogadro's number. There are $0.5ZN_A$ bonds per mole.

If we cleave a face-centered cubic crystal along a (111) plane, three bonds per atom will be broken. Because there are two surfaces formed, the work required to form the surfaces will be $\frac{3}{2}\varepsilon$ per surface atom. Combining with Eq. 4.2, the work per surface atom is

$$w = \tfrac{3}{2}\varepsilon = \frac{\Delta \underline{H}_s}{4N_A} \quad (\text{for } Z = 12)$$

The surface energy γ is then

$$\gamma = \frac{\Delta H_s}{4N_A}\left(\frac{N}{A}\right) \qquad (4.3)$$

where the term N/A is the number of atoms per unit area. For a face-centered cubic structure, N/A is $4/(\sqrt{3}a_0^2)$ where a_0 is the lattice parameter. For copper, the enthalpy of sublimation is approximately 170,000 J/(g·mol), and the lattice spacing, a_0, is 3.615 Å. The calculated work of forming a free surface of copper is approximately 1400 ergs/cm². The measured value is about 1600 ergs/cm². As we expected, the simple model does not yield accurate results because the approximations used do not account for other changes that take place in the vicinity of the surface. However, the result is not unreasonable.

An important conclusion to be drawn from this simple calculation is that the surface energy of a solid depends on the crystallographic plane. The work required to create the surface, γ, depends on the number of bonds broken per atom when the surface is created, and the number of atoms per area of surface. For a (111) plane, there are three bonds broken per atom, and the number of atoms per unit area is (Figure 4.2):

$$\left(\frac{N}{A}\right)_{(111)} = \frac{4}{a_0^2\sqrt{3}} \qquad (4.4)$$

For a (100) plane there are four broken bonds per atom, and the number of atoms per unit area (Figure 4.3) is:

$$\left(\frac{N}{A}\right)_{(100)} = \frac{2}{a_0^2} \qquad (4.5)$$

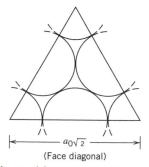

$a_0\sqrt{2}$
(Face diagonal)

Figure 4.2 Atomic packing on the (111) face of a face-centered cubic crystal.

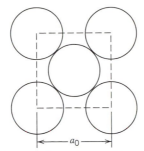

Figure 4.3 Atomic packing on the (100) face of a face-centered cubic crystal.

The ratio of the two surface energies is, thus, about

$$\frac{\gamma_{(100)}}{\gamma_{(111)}} = \frac{2}{\sqrt{3}} = 1.15 \tag{4.6}$$

The foregoing is not intended as an accurate calculation of these surface energies, but simply as an illustration that surface energies of different crystallographic faces are not expected to be equivalent. In solids the surface energy is not isotropic. It is a function of crystallographic orientation. As a consequence, the equilibrium shape of a single crystal is not a sphere. Its equilibrium shape depends on the relative surface energies of the crystallographic planes, as will be demonstrated in Section 4.13.

4.3 EFFECT OF SURFACE CURVATURE

In physical systems, the surface between two phases is not an infinitely thin, geometrical plane. The disturbances at a surface extend several atomic layers into each of the two phases surrounding the surface. But, in a system with only one component, C, we can locate a geometrical surface between two phases in a way that assigns no mass to the surface. Figure 4.4 shows the concentration of component C as a function of distance near the interface between phases I and II. The surface between these two phases, indicated by y, will be assigned no mass if the shaded area to the left of the surface is equal to the shaded area to the right. The entire mass in the system will be accounted for by the phases I and II, with no mass left over to be accounted for by the surface ($m_y = 0$). Note that the assignment of no mass to a surface, which is possible in one-component systems, is not, in general, possible for systems containing more than one component (Section 4.11).

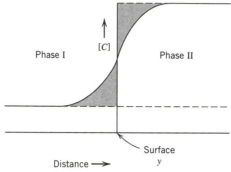

Figure 4.4 Concentration of component *C* as
a function of distance. Surface is assumed to
be at location where shaded areas are equal.

For a single-component system, consider phases I and II separated by a surface
S (Figure 4.5). If this surface moves from *A* to *B*, the volumes (and masses) of the
phases will change: phase I increasing and phase II decreasing. The total change of
Gibbs free energy can be written as follows:

$$dG = \underline{G}_I \, dm_I + \underline{G}_{II} \, dm_{II} + \gamma \, dA = \delta w_{rev} \qquad (4.7)$$

where dm_I and dm_{II} are the changes in the masses (measured in moles) of phases I
and II as the infinitesimal movement of the surface takes place

At equilibrium the reversible work (Eq. 4.7) is zero. The partial molar Gibbs free
energy (the chemical potential, μ) for a single-component system is the same as the
specific Gibbs free energy $(\overline{G} = \mu = \underline{G})$

$$\mu_I \, dm_I + \mu_{II} \, dm_{II} + \gamma \, dA = 0 \qquad (4.8)$$

Noting that the increase in mass of phase I is equal to the decrease of mass of
phase II $(dm_I = -dm_{II} = dm)$,

$$\mu_I - \mu_{II} = \gamma \, \frac{dA}{dm} \qquad (4.9)$$

Figure 4.5 Movement of a surface with infinite radius of curvature (flat).

If the surface is *flat* (infinite radius of curvature), then dA/dm is zero, because the area of the interface does not change as mass moves from I to II. Thus, the chemical potential of phase I is equal to the chemical potential of phase II. This is sometimes written as follows:

$$\mu_I = \mu_{II} = \mu_\infty \tag{4.10}$$

where μ_∞ is the chemical potential of a material with an infinite radius of curvature.

For a *curved* surface (Figure 4.6) Eq. 4.10 is not true. The area of the surface changes as the boundary between phases II and I moves. Introducing the molar volume ($\underline{V} = dV/dm$) into Eq. 4.9, we have

$$\mu_I - \mu_{II} = \gamma \frac{dA}{dm} = \underline{V}\gamma \frac{dA}{dV} \tag{4.11}$$

The ratio dA/dV is not zero for a curved surface. For a spherical surface:

$$V = \frac{\omega r^3}{3} \quad \text{and} \quad A = \omega r^2$$

where ω is the solid angle subtended by the portion of the surface being considered:

$$dV = \omega r^2 dr \quad \text{and} \quad dA = 2\omega r\, dr$$

$$\frac{dA}{dV} = \frac{2}{r}$$

Therefore,

$$\mu_I - \mu_{II} = \underline{V}\frac{2\gamma}{r} \tag{4.12}$$

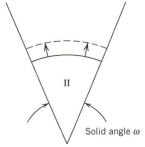

Figure 4.6 Movement of a spherical surface with finite radius of curvature.

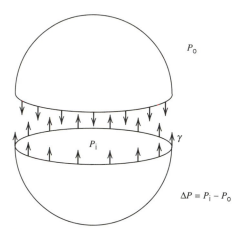

Figure 4.7 Force holding two hemispheres together is $2\pi r\gamma$; force pushing them apart is $\pi r^2 \Delta P$, where $\Delta P = 2\gamma/r$.

where $\underline{V}$ is the molar volume and γ is the interfacial (surface) energy between phases I and II.

Now let us compare the two conditions. If condition II is equivalent to a material with an infinite radius of curvature, and condition I entails a finite radius of curvature r, we can write:

$$\mu_r - \mu_\infty = \underline{V}\frac{2\gamma}{r} \tag{4.13}$$

Thus the chemical potential of a material with a radius of curvature r is greater than the chemical potential for that same material with an infinite radius of curvature. This difference in chemical potential manifests itself in several ways, which are discussed in Sections 4.4, 4.5, and 4.6.

For a nonspherical, curved surface with principal radii of curvature r_1 and r_2, the difference in chemical potential is[2]:

$$\mu_{r_1,r_2} - \mu_\infty = V\gamma \left(\frac{1}{r_1} + \frac{1}{r_2}\right) \tag{4.14}$$

---

[2]In publications concerning surface effects, one often finds the equation for the difference in chemical potential written in terms of the mean curvature of the surface κ, defined as follows:

$$\kappa = \frac{1}{2}\left(\frac{1}{r_1} + \frac{1}{r_2}\right)$$

The form of Eq. 4.13 then becomes:

$$\mu_r - \mu_\infty = 2\underline{V}\gamma\kappa$$

Equation 4.13 can also be derived using a mechanical argument combined with thermodynamics. If we think of the surface as a membrane surrounding the condensed phase, and the surface tension as the stress in the membrane, we can establish the pressure difference across the curved surface through a force balance (Figure 4.7). The force tending to push the two halves of the sphere apart is the product of the cross-sectional area of the sphere at its mid-point (πr^2) and the difference in pressure between the inside and the outside of the sphere ($P_i - P_o = \Delta P$). This force, $\pi r^2 \Delta P$, is balanced by the surface tension force, $2\pi r\gamma$. Thus:

$$\pi r^2 \, \Delta P = 2\pi r\gamma \tag{4.15}$$

$$\Delta P = \frac{2\gamma}{r}$$

This increased pressure on the condensed phase translates into a difference in chemical potential.

$$d\mu = \underline{V} \, dP \text{ (at constant } T)$$

$$\int_{\mu_0}^{\mu_i} d\mu = \underline{V} \int_{P_0}^{P_i} dP = \underline{V}(P_i - P_0) = \underline{V} \, \Delta P$$

$$\mu_r - \mu_\infty = \underline{V} \frac{2\gamma}{r} \tag{4.13}$$

4.4 VAPOR PRESSURE

If a material in the condensed state (either solid or liquid) is in equilibrium with its vapor, the chemical potential of the condensed material μ_c is equal to the chemical potential of its vapor μ_v:

$$\mu_c = \mu_v$$

The equation above is valid for both curved surfaces and flat surfaces (infinite radius of curvature)

$$\mu_{c,r} = \mu_{v,r}$$

$$\mu_{c,\infty} = \mu_{v,\infty}$$

where $\mu_{c,r}$ is the chemical potential of the condensed material with radius of curvature r, and $\mu_{v,r}$ is the chemical potential of the vapor in equilibrium with it.

From Eq. 4.13,

$$\mu_{c,r} - \mu_{c,\infty} = \frac{V2\gamma}{r}$$

From the definition of fugacity:

$$\mu_{c,r} - \mu_{c,\infty} = RT \ln \frac{f_r}{f_\infty}$$

Because the vapor in each case is in equilibrium with the condensed material:

$$\mu_{v,r} - \mu_{v,\infty} = RT \ln \frac{f_r}{f_\infty}$$

If the vapor behaves like an ideal gas:

$$\mu_{v,r} - \mu_{v,\infty} = RT \ln \frac{P_r^\circ}{P_\infty^\circ}$$

where P_r° is the vapor pressure of a particle of radius r.
Therefore:

$$\ln \frac{P_r^\circ}{P_\infty^\circ} = \frac{V}{RT}\frac{2\gamma}{r} \qquad (4.16)$$

Thus the vapor pressure of a spherical particle is a function of its radius. Smaller particles (with smaller radii of curvature) have higher vapor pressures. Therefore, if a material exists in a variety of particle sizes, the smaller particles will tend to shrink and the larger particles will tend to grow because the vapor pressure of the material above the smaller particles will be greater than the vapor pressure above the larger particles. The gas molecules will diffuse from the vicinity of the smaller particles to the larger ones.

4.5   SOLUBILITY OF SMALL PARTICLES

Another consequence of the variation of chemical potential with radius of curvature can be observed in the solubility of particles of various sizes. For the two-component system illustrated in Figure 4.8, a plot of activity versus concentration is shown below the phase diagram. For the sake of simplification, assume that the solubility of A in B is vanishingly small. The phase β is taken to be pure B.

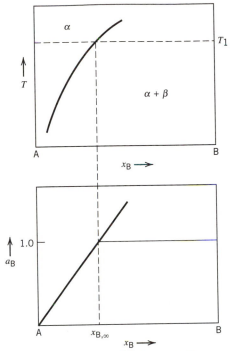

Figure 4.8 Activity of B in an A–B system
with no solubility of A in B.

If material B exists as small spherical particles, the chemical potential of B as a
function of its radius of curvature (from Eq. 4.13) is:

$$\mu_{B,r} - \mu_{B,\infty} = \frac{2V\gamma_{\alpha-B}}{r} \tag{4.17}$$

The surface energy, $\gamma_{\alpha-B}$, in this equation is the interfacial energy between pure
B and phase α.

Define the standard state for B as pure B with an infinite radius of curvature.
From Chapter 1 (Eqs. 1.49 and 1.50), we can write

$$\mu_{B,r} = \mu_{B,\infty} + RT \ln a_B \tag{4.18}$$

If the activity of B is linearly proportional to its mole fraction in phase α (Henry's
law), then:

$$a_B = kx_B \tag{4.19}$$

   The activity of B is one at the phase boundary between the α and $\alpha + B$ fields. The composition at this point is $x_{B,\infty}$, the solubility of B in A for particles with flat surfaces (infinite radius of curvature).

$$1 = kx_{B,\infty} \qquad \text{or} \qquad k = \frac{1}{x_{B,\infty}} \tag{4.20}$$

Combining Eqs. 4.18, 4.19, and 4.20, we have

$$\mu_{B,r} - \mu_{B,\infty} = RT \ln \frac{x_{B,r}}{x_{B,\infty}} \tag{4.21}$$

Combining with Eq. 4.17:

$$\ln \frac{x_{B,r}}{x_{B,\infty}} = \frac{V}{RT} \frac{2\gamma_{\alpha-B}}{r} \tag{4.22}$$

where $x_{B,r}$ is the solubility of B in A (as phase α) when the radius of curvature of the B particles is r, and $x_{B,\infty}$ is the solubility of B in α when phase B has an infinite radius of curvature.

   The solubility of B *increases* as its particle size (radius of curvature) *decreases*. If, as in the case illustrated in Figure 4.9, phase B exists as small spheres and large spheres, the solubility of B will vary with particle size. In the material immediately surrounding the small particles, the concentration of B will be higher. If the sample is held at a high temperature where B atoms are mobile, the small particles will tend to dissolve and the larger particles will grow. The process is known as *coarsening* or *Ostwald ripening*.

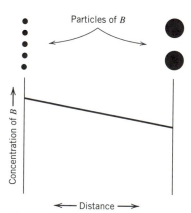

Figure 4.9 Solubility as a function of particle size.

4.6 MELTING TEMPERATURE OF SMALL PARTICLES

In the preceding sections we showed that the curved surface of a particle influences its vapor pressure and its solubility. In this section we consider the melting point of small, spherical particles, and we find that their melting temperature depends on their radii. The same conclusion applies, of course, to surfaces of a solid with different radii of curvature. The melting temperature of the point of a needle is different from the melting point of the body of the needle, a concept useful in the study of solidification of materials. To return to the question of the melting temperature of small, spherical particles, consider a particle of solid (subscript s) in equilibrium with its liquid (subscript l). The equations of equilibrium are:

$$T_s = T_l$$

$$\mu_s = \mu_l$$

The pressures in solid and liquid are related by:

$$P_s = P_l + \frac{2\gamma_{l-s}}{r} \tag{4.23}$$

Based on the definition of the chemical potential:

$$d\mu_l = -\underline{S}_l dT + \underline{V}_l dP_l$$

$$d\mu_s = -\underline{S}_s dT + \underline{V}_s dP_s = -\underline{S}_s dT + \underline{V}_s d\left(P_l + \frac{2\gamma_{l-s}}{r}\right)$$

At equilibrium, the chemical potentials of solid and liquid are equal, and they remain equal as temperature and pressure change. Thus, their differentials are equal.

$$-\underline{S}_s dT + \underline{V}_s d\left(P_l + \frac{2\gamma_{l-s}}{r}\right) = -\underline{S}_l dT + \underline{V}_l dP_l$$

$$(\underline{S}_l - \underline{S}_s)dT - (\underline{V}_l - \underline{V}_s)dP_l - 2\underline{V}_s\gamma_{l-s}\frac{dr}{r^2} = 0 \tag{4.24}$$

Assuming that we are not concerned with overall pressure changes on the liquid ($dP_l = 0$), and noting that $\underline{S}_l - \underline{S}_s = \Delta\underline{S}_m$, the entropy of melting, we have

$$\Delta\underline{S}_m dT = 2\underline{V}_s\gamma_{l-s}\frac{dr}{r^2} \tag{4.25}$$

Integrating from the melting temperature of a large particle (with infinite radius of curvature) to the melting temperature of a particle with radius r (assuming that the entropy of melting is constant over a small temperature range), we have

$$\Delta \underline{S}_m \int_{T_\infty}^{T_r} dT = 2\underline{V}_s \gamma_{l-s} \int_{r=\infty}^{r} \frac{dr}{r^2}$$

$$\Delta \underline{S}_m (T_r - T_\infty) = \Delta \underline{S}_m \Delta T = - \frac{2\underline{V}_s \gamma_{l-s}}{r} \qquad (4.26)$$

$$\Delta T = - \frac{2\underline{V}_s \gamma_{l-s}}{\Delta \underline{S}_m} \frac{1}{r}$$

The entropy of melting is the enthalpy of melting divided by the "equilibrium" melting temperature, that is, the melting temperature of large particles, thus:

$$\Delta T = - \frac{2\underline{V}_s \gamma_{l-s} T_m}{\Delta \underline{H}_m} \frac{1}{r} \qquad (4.27)$$

Based on the derivation above, we conclude that the change in melting temperature of a small, spherical particle is inversely proportional to its radius. The same conclusion applies to surfaces. The melting temperature is inversely proportional to the radius of curvature (or the sum of the inverse principal radii of curvature).

To estimate the magnitude of this effect, let us calculate the melting temperature of very fine particles of gold, with radius 100 Å (0.01 μm) compared to the melting temperature of bulk gold. For gold:

$$T_m = 1336 \text{ K}$$

$$\Delta \underline{H}_m = 12{,}360 \text{ J/mol}$$

$$\underline{V} = 10.2 \times 10^{-6} \text{ m}^3/\text{mol}$$

$$\gamma_{l-s} = 132 \text{ ergs/cm}^2 = 0.132 \text{ J/m}^3$$

$$\Delta T = - \frac{2(0.132)(10.2 \times 10^{-6})(1336)}{12{,}360} \times \frac{1}{10^{-8}}$$

$$\Delta T = T_r - T_\infty = -29.1 \text{ K}$$

This is a large difference in melting temperature. But, of course, 0.01 μm diameter is a very small particle size.

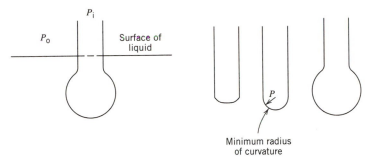

Figure 4.10 Bubble method for surface energy determination.

4.7 MEASUREMENT OF SURFACE ENERGY

In liquids, surface energy can be measured by introducing bubbles of an inert gas into the liquid through a capillary tube. Neglecting height differences, the pressure inside the capillary tube (Figure 4.10) and outside is equal to:

$$\Delta P = P_i - P_0 = \frac{2\gamma}{r} \tag{4.28}$$

As bubbles are produced, the radius of curvature goes through a minimum when the radius of the bubble is exactly equal to the radius of the capillary tube (Figure 4.10). From Eq. 4.28, ΔP goes through a maximum when r is at a minimum. By measuring the maximum pressure and knowing the radius of the capillary tube, we can find the value of the surface energy.

Another method involves the rise of a liquid on the inside of a capillary tube. In the experiment illustrated in Figure 4.11, the surface energy can be determined by

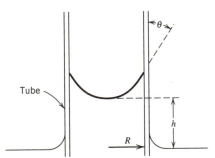

Figure 4.11 Capillary rise method for surface energy measurement.

knowing the height of rise in a capillary, the radius of the capillary, and the angle between the liquid inside the capillary and the walls.

$$\Delta P = \rho g h = \frac{(2 \cos \theta)\gamma}{R}$$

$$\gamma = \frac{R \rho g h}{\cos \theta}$$

(4.29)

where ρ is the density of the fluid and R is the radius of the capillary.

The measurement of the surface energy of liquids requires careful experimental technique because any contamination of the surface will change the results. Two types of measurement have been described in this section. Many other techniques are also available (Ref. 1). The measurement of liquid surface energy (surface tension) is more easily accomplished than the measurement of the surface energy of solid, which we discuss next.

4.8 SURFACE ENERGY OF SOLIDS

An interesting technique has been used to measure the surface energy of solids (Ref. 2). Wires of the solid (all of the same diameter) are suspended from a plate with weights placed at the bottom of each (Figure 4.12). The wires are then heated to a high temperature, about 0.9 of the melting point on an absolute scale. At that temperature atomic mobility is high and the surfaces have a liquidlike quality. The wires will change dimension because of two forces. They will tend to elongate because of the stress imposed by the weights. They will tend to shrink because of the surface forces, which try to minimize the surface area. The resulting elongation (or contrac-

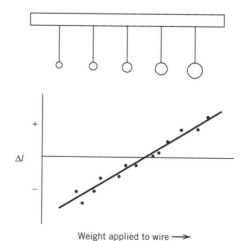

Figure 4.12 Elongation method for measurement of surface energy of solids.

tion) is plotted as a function of the applied weight to establish the balance point (zero elongation). A consideration of the energy balance on the wires leads to:

$$\delta w_{rev} = mg\, dl = \gamma\, dA \qquad (4.30)$$

The surface area, A, of the wires changes because of changes in radius and length:

$$A = 2\pi r l \qquad (4.31)$$

$$dA = 2\pi(r\, dl + l\, dr)$$

But, the volume of the sample (V) is constant.

$$V = \pi r^2 l$$

$$dV = 2\pi l r\, dr + \pi r^2 dl = 0 \qquad (4.32)$$

$$dr = -\frac{1}{2}\frac{r}{l}\, dl$$

Combining Eqs. 4.30, 4.31, and 4.32 for the external surface at the balance point (no elongation), we have

$$\gamma = \frac{mg}{\pi r} \qquad (4.33)$$

Solids, when compared to liquids, have another complicating feature, internal surfaces such as grain boundaries. In the samples described in Ref. 2, the grain boundaries were found to be located and oriented in the manner illustrated in Figure 4.13. As the sample was stretched, the change in the surface grain boundary energy could be expressed as

$$-\delta w = dG_{gb} = \gamma_{gb}d(A_{gb}) \qquad (4.34)$$

where γ_{gb} is the grain boundary energy, A_{gb} is the total grain boundary area, and

$$dG_{gb} = \gamma_{gb}d(n\pi r^2) \qquad (4.35)$$

where n is the number of grain boundaries per unit length.
   Combining Eqs. 4.33 and 4.35 yields

$$mg = \pi\gamma r - \pi r^2 \frac{n}{l}\gamma_{gb} \qquad (4.36)$$

This experimental method is interesting because one can obtain not only the surface energy of the material against its own vapor, but also the surface energy against a variety of gases, simply by varying the atmosphere in the furnace in which

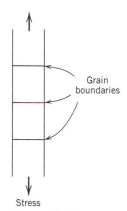

Figure 4.13 Location of grain
boundaries in elongation method.

the experiment is conducted. The method does, however, have limitations. Experiments must be conducted at temperatures high enough for the atoms to be mobile, that is, close to the melting point. The technique also yields a *single* surface energy value for a solid–vapor combination, and, as discussed in Section 4.2, the surface energy of a solid is dependent on the crystallographic plane exposed. Nevertheless, the zero elongation technique yields some useful average values for the surface energy of solids.

4.9   RELATIVE SURFACE ENERGIES

The relative surface energies among solid phases can be determined by measuring the angles among them after the phases have been allowed to equilibrate. Consider three phases of crystals of different orientation (labeled 1, 2, and 3) separated by

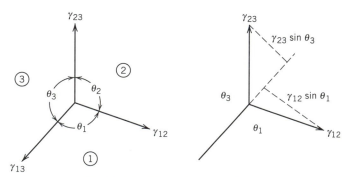

Figure 4.14 Three-phase equilibrium.

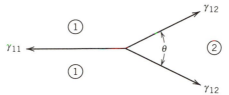

Figure 4.15 Two-phase equilibrium.

three surfaces, 1 2, 1 3, 2 3 (Figure 4.14). Assuming that the surface were free to move, the force balance at equilibrium yields:

$$\gamma_{23} \sin \theta_3 = \gamma_{12} \sin \theta_1 \qquad (4.37)$$

From this force balance, it is possible to show that the relative surface energies can be determined to be

$$\frac{\gamma_{23}}{\sin \theta_1} = \frac{\gamma_{12}}{\sin \theta_3} = \frac{\gamma_{31}}{\sin \theta_2} \qquad (4.38)$$

If two of the phases are identical, then (Figure 4.15)

$$\gamma_{11} = 2\gamma_{12} \cos \left(\frac{\theta}{2}\right) \qquad (4.39)$$

This latter case is important in the determination of grain boundary energies through thermal etching or grooving. If a solid material with grain boundaries is heated for a long time in a vacuum furnace, it will develop grooves where the grain boundaries intersect the free surface. The groove angle at the surface is a measure of the grain boundary energy relative to the surface energy of the solid itself. Using Eq. 4.39, γ_{11} is the grain boundary energy, γ_{12} is the surface energy of the solid against its vapor, and θ is the groove angle.

4.10   WETTING OF SURFACES

The interfacial energy between a solid and a liquid can be inferred by measuring the contact angle between the solid surface and a droplet of the liquid in question, as in Figure 4.16. This is called the sessile drop technique. The solid with the liquid drop on it is heated to the temperature of interest in a furnace with one clear end, permitting the shape of the droplet and the contact angle to be observed using a telescope. We assume that the surface is not free to move in the vertical direction, so

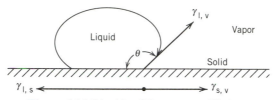

Figure 4.16 Liquid–solid–vapor equilibrium.

that a force balance in that direction has no meaning. A force balance in the horizontal direction yields:

$$\gamma_{l,s} = \gamma_{s,v} - \gamma_{l,v} \cos \theta \qquad (4.40)$$

Based on Eq. 4.40, the solid–liquid interfacial energy $\gamma_{l,s}$ can be determined if θ is measured and the solid and liquid surface energies, $\gamma_{s,v}$ and $\gamma_{l,v}$, are known.[3] In general, the liquid is said to wet the solid if the angle θ is less than 90°. Values of θ greater than 90° are called nonwetting. Of course, at $\theta = 180°$, the solid is not wet at all.

In many cases, such as the application of a glaze in ceramics, we are interested in complete wetting, that is, spreading of a liquid on a solid surface. If a liquid is to spread on a solid, then to make the angle θ in Eq. 4.40 zero, $\gamma_{s,v}$ must be greater than the sum of $\gamma_{l,s}$ and $\gamma_{l,v}$.

4.11   SURFACE ENERGY AND IMPURITY SEGREGATION AT INTERFACES (GIBBS ADSORPTION ISOTHERM)

Figure 4.17 shows two phases, α and β, separated by a surface, y. The α and β phases each consist of A and B atoms. Consider the material A to be the solvent, and B the solute. In this section we explore the criteria for impurity (solute) segregation at the interface—that is, concentrations of solute of the interface in excess of the concentration in the bulk α and β phases.

For a two-component system with temperature and pressure constant, we can write an expression for the Gibbs free energy at the interface:

$$G^y = \gamma A + m_A^y \mu_A^y + m_B^y \mu_B^y \qquad (4.41)$$

$$dG^y = \gamma \, dA + A \, d\gamma + m_A^y d\mu_A^y + \mu_A^y dm_A^y + m_B^y d\mu_B^y + \mu_B^y dm_B^y \qquad (4.42)$$

---

[3]The shape of the top portion of the droplet can actually be used to measure the surface energy of the liquid (Ref. 3, Section 8.2).

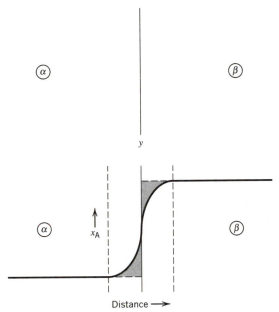

Figure 4.17 Concentrations at interface.

Proceeding in a manner used to derive the Gibbs–Duhem equation (Section 1.7.5),

$$\Sigma \mu_i dm_i + \gamma \, dA = 0 \tag{4.43}$$

$$dG^y = \gamma \, dA + m_A^y d\mu_A^y + m_B^y d\mu_B^y \tag{4.44}$$

At equilibrium, $dG^y = 0$. To simplify Eq. 4.44, consider a mass balance for the materials involved.

The mass balance for B is:

$$m_B = x_B^\alpha m_\alpha + x_B^\beta m_\beta + m_B^y \tag{4.45}$$

where m_B = total mass of B
m_α, m_β = mass of α, mass of β
x_B^α = mole fraction of B in α
m_β^y = mass of β at interface

We define the surface excess concentration of B at y as:

$$\Gamma_B^y = \frac{m_B^y}{A} \tag{4.46}$$

Then:

$$m_B = x_B^{\alpha} m_{\alpha} + x_B^{\beta} m_{\beta} + \Gamma_B^y A \tag{4.47}$$

A similar mass balance for A is:

$$m_A = x_A^{\alpha} m_{\alpha} + x_A^{\beta} m_{\beta} + \Gamma_A^y A \tag{4.48}$$

As noted in the introduction to this chapter, the physical interface between phases α and β is not a sharp geometrical surface. The compositions of α and β do not change discontinuously at y. The concentration of A (x_A) across the interface region is shown schematically in Figure 4.17.

For the purpose of our analysis, we may locate the geometrical interface (y) at any place we choose. Let us locate y, as we did in Section 4.3, at a place where there is no excess concentration of A at y (i.e., where the two shaded areas in Figures 4.4 and 4.17 are equal). Note that we can do this for material A, but not, in general, for material B. Because all of material A can then be accounted for in either phase α or β, there is no excess surface concentration of A. Thus:

$$\Gamma_A^y = 0 \tag{4.49}$$

and

$$m_A = x_A^{\alpha} m_{\alpha} + x_A^{\beta} m_{\beta} \tag{4.50}$$

We can now explore the factors that influence the concentrations of B at the interface, Γ_B^y.

Dividing Eq. 4.44 by A yields

$$d\gamma + \Gamma_A^y \, d\mu_A^y + \Gamma_B^y \, d\mu_B^y = 0 \tag{4.51}$$

By our definition $\Gamma_A^y = 0$, hence:

$$d\gamma = - \Gamma_B^y \, d\mu_B \tag{4.52}$$

But, at equilibrium $\mu_B^y = \mu_B^{\alpha} = \mu_B^0 = \mu_B$:

$$d\gamma = -\Gamma_B^y \, d\mu_B \tag{4.53}$$

$$\Gamma_B^y = - \left(\frac{\partial \gamma}{\partial \mu_B} \right)_T$$

But,

$$\mu_B = \mu_B^0 + RT \ln a_B \tag{4.54}$$

$$d\mu_B = RT \, d \ln a_B$$

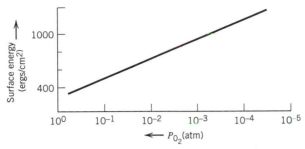

Figure 4.18 Surface energy as a function of activity (pressure) of oxygen.

In regions where the activity coefficient of B is constant, we may write:

$$d\mu_B = RT \, d \ln x_B \tag{4.55}$$

Hence:

$$\Gamma_B^\gamma = -\frac{1}{RT}\left(\frac{\partial \gamma}{\partial x_B}\right)_T \tag{4.56}$$

From Eq. 4.56 we conclude that if the addition of B to a system *lowers* the surface energy of the α–β interface, material B will *segregate* preferentially at the interface. If one of the phases (e.g., β) is a vapor phase, material B will be adsorbed on the α surface.

An example of this is the measurement made on the surface energy of silver as a function of oxygen pressure as shown in Figure 4.18. As the pressure of oxygen increases (to the left in the figure), the surface energy decreases. This means that oxygen segregates preferentially (adsorbs) on the silver surface, the interface between the two phases.

4.12 ADSORPTION ON SOLIDS (ADSORPTION ISOTHERMS)

The Gibbs adsorption isotherm, discussed in Section 4.11, relates the adsorption behavior of a component to its effect on the surface energy of the solid on which it is adsorbed. There are other relationships that are important in the study of adsorption. For example, in the case of gas adsorption on solids it is important to know the relation between the pressure of an adsorbing gas and the fraction of a solid surface occupied by the adsorbate.

The nature of gas adsorption on solids is divided broadly into two classes, physical adsorption and chemisorption. Physical adsorption results from relatively weak van der Waals forces, where the bonding energy is of the order of a few kilojoules per mole. Multiple adsorption layers are possible in physical adsorption. An example is the adsorption of nitrogen on iron at low temperatures, about 77 K. In chemisorption, the bonding between the gas and the solid is quite strong, with bonding energies of tens or hundreds of kilojoules per mole. In this case, the adsorption layer is very tightly bound to the solid, and the adsorbing species form monolayers. An example is the adsorption of oxygen on silica surfaces.

One description of an adsorption isotherm is due to Langmuir, who derived the relationship between the degree of surface coverage ϕ and the pressure of the adsorbing gas P, using a kinetic argument. The degree of coverage ϕ is the number of occupied surface sites N_O divided by the total number of surface sites N_S. Langmuir reasoned that the rate of *desorption* from the surface $\dot{m}_d$ is proportional to the fraction of surface sites *occupied* by the adsorbate:

$$\dot{m}_d = k_d\, \phi \tag{4.57a}$$

The rate of *adsorption* $\dot{m}_a$ is proportional to the pressure of the adsorbate and the fraction of sites *unoccupied:*

$$\dot{m}_a = k_a P(1 - \phi) \tag{4.57b}$$

At equilibrium, the rate of adsorption is equal to the rate of desorption, hence:

$$k_d\, \phi = k_a P(1 - \phi)$$

This is usually written as follows:

$$\phi = \frac{KP}{KP + 1} \tag{4.58}$$

where $K = k_a/k_d$.

Figure 4.19 plots ϕ versus P. The Langmuir adsorption isotherm is useful in describing chemisorption because it accounts for the formation of only one adsorption layer. Other analyses, such as the Brunauer–Emmet–Teller (BET) isotherm, are

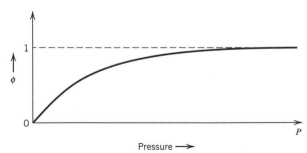

Figure 4.19 Langmuir adsorption isotherm: fraction of surface covered versus pressure of adsorbing gas.

used to relate θ to P in physical adsorption, where more than one layer of adsorbed species are observed (Ref. 1).

4.13 EQUILIBRIUM SHAPE OF A CRYSTAL

If the surface energy of a condensed phase is isotropic, as in a liquid, then, based on common experience, the condensed phase will assume a spherical shape at equilibrium. If, however, the surface energy varies with crystallographic orientation, as in a solid, one would expect the equilibrium shape of a solid crystal to be determined by the relative surface energies of the crystallographic planes. The presence of low surface energy planes would be favored, and the equilibrium shape would not be a sphere. This has been confirmed experimentally. When single-crystal spheres are etched or partially vaporized, the sphere becomes a faceted solid in which certain planes are favored[4] (Refs 4 and 5).

The equilibrium shape of a crystal is determined by minimizing its surface free energy. If a crystal may be found by planes 1, 2, 3, . . . , whose exposed area will be $A_1, A_2, A_3, \ldots$, the equilibrium shape will be the one that minimizes $\Sigma \gamma_i A_i$. A graphical method for doing this uses a Wulff plot (Ref. 6). In this method, the surface energy of a crystal is plotted in polar coordinates. Planes are drawn at each point normal to the radius. The equilibrium shape of the crystal will be defined by the interior surface of the planes erected perpendicular to the polar plot. This is illustrated schematically in two dimensions in Figure 4.20.

The development of facets in two dimensions can be illustrated by a simple example. Consider the two-dimensional, cubic crystal in Figure 4.21a. We are look-

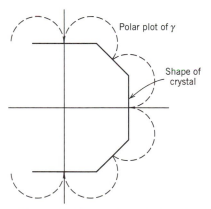

Figure 4.20 Example of Wulff plot.

---

[4]The facets that appear when single crystals are etched may be due to kinetic effects as well as to equilibrium considerations. The rate of etching may vary with crystallographic orientation.

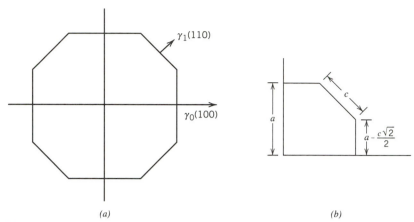

(a)   (b)

Figure 4.21 (*a*) Two-dimensional cubic crystal. (*b*) Upper right-hand quadrant of crystal in (*a*).

ing in the [100] direction. Assume that either of two planes (but no others) may constitute the exterior surface, either the (100), the (110), or combinations of both. We will calculate the "*a*" and "*c*" dimensions of the crystal by minimizing the surface energy E_S at constant crystal area, A. The crystal is symmetrical in the vertical and horizontal directions; hence we may isolate the upper right hand quadrant for our analysis (Figure 4.21*b*).

$$E_S = 2 \left(a - \frac{c\sqrt{2}}{2} \right) \gamma_0 + c\gamma_1 \tag{4.59}$$

$$A = a^2 - \frac{c^2}{4} \tag{4.60}$$

We minimize the surface energy be setting the differential of E_S with respect to "*c*" equal to zero at constant A. The result is:

$$\frac{c}{a} = 2 \left(\sqrt{2} - \frac{\gamma_1}{\gamma_0} \right) \tag{4.61}$$

The ratio *c/a* is a measure of the relative amounts of exposed (110) and (100) planes. The amount of (110) is proportional to "*c*". The amount of (100) is proportional to "$a - c\sqrt{2}/2$". If γ_1 is equal to or greater than $\sqrt{2}\gamma_0$, then c will be zero; that is, the (110) surface will not be exposed at equilibrium. If the surface energies of the planes are equal, $\gamma_0 = \gamma_1$, then:

$$2 \left(a - \frac{c\sqrt{2}}{2} \right) = c \tag{4.62}$$

and the size of the two facets will be equal. This simplified example illustrates how various crystal faces can be exposed at equilibrium, the face with the lower energy being favored.

4.14   EFFECT OF TEMPERATURE ON SURFACE ENERGY

We saw in Section 4.2 that surface atoms have fewer bonds to their nearest neighbors than atoms away from the surface. The surface atoms, consequently, have higher energies. One would also expect them to differ entropically from atoms in the bulk of the solid. There should, therefore, be an entropy associated with the surface. This surface entropy is related to the temperature coefficient of surface energy. The relationship can be derived by considering the differential of the Gibbs free energy at constant pressure:

$$dG = -SdT + \gamma \, dA$$

Using the cross-differentiation technique (Chapter 1, Eq. 1.29), we write

$$\left(\frac{\partial S}{\partial A}\right)_{T,P} = -\left(\frac{\partial \gamma}{\partial T}\right)_{A,P} = S_s \tag{4.63}$$

where S_s is the entropy per unit surface, the surface entropy.

The surface entropy is generally positive; hence the temperature coefficient of surface energy is usually negative.

REFERENCES

1. Adamson, Arthur W., *Physical Chemistry of Surface,* 5th ed. Wiley Interscience, New York, 1990, Chapter 2.
2. Buttner, F. H., Funk, E. R., and Udin, H., *J. Phys. Chem., 56,* 657 (1952).
3. Lupis, C. H. P., *Chemical Thermodynamics of Materials,* North Holland, Amsterdam, 1983.
4. Batterman, B. W., *J. Appl. Phys. 28,* 1236 (1957).
5. Gatos, H. C., et al., *Fundamental Phenomena in Materials Science,* Vol. 2, Plenum Press, New York, 1966.
6. Wulff, G., *Z. Kristallogr., 34,* 449 (1901).

PROBLEMS

4.1 In an experiment, liquid mercury is to be introduced into a cylindrical hole of 1 μm *diameter* in a material, A. The other side of the hole is open to the atmosphere (i.e., the pressure is 1 atm).

(a) What is the *total* pressure that must be applied to the mercury to introduce it into the hole if the mercury does not wet material A at all (i.e., the wetting angle is 180°)?

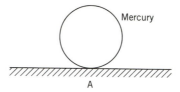

(b) What is the *total* pressure that must be applied if the wetting angle of mercury on material A is 90°?

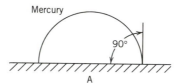

DATA

The interfacial energy of mercury–air is 487 ergs/cm².

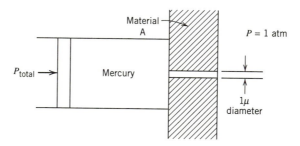

4.2 A nickel wire 0.005 in. in diameter is suspended in its own vapor at 1300°C. It is found that a weight of 0.0366 g is required to balance the tendency of the wire to shrink, and there are 30 grains per centimeter of wire length. Calculate the surface energy (assuming that $\gamma gb = 1/3\gamma sv$).

4.3 A liquid silicate with surface tension of 500 ergs/cm² makes contact with a polycrystalline oxide with an angle $\theta = 45°$ on the surface of the oxide. If mixed with the oxide, it forms liquid globules at three grain intersections. The average dihedral angle ϕ is 90°. If we assume the interfacial tension of the oxide–oxide interface, without the silicate liquid is 1000 dynes/cm, compute the surface tension of the oxide.

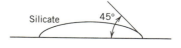

4.4 At a high temperature (about 2000 K), the surface energy of alumina, Al_2O_3, is 900 ergs/cm². For liquid iron against its own vapor, the surface energy is 1700 ergs/cm². Under the same conditions, the interfacial energy between iron and alumina is about 2300 ergs/cm². What will be the contact angle of a small piece of iron melted on an alumina plate?

	Melting Temperatures (K)
Alumina	2300
Iron	1809

4.5 One way in which an adhesive forms a bond is illustrated in the accompanying figure.

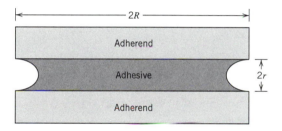

An adhesive that wets two solids will form a thin film with a concave meniscus at the liquid–air interface. The area of contact of the adhesive with the adherend is circular, with radius R and thickness $2r$. The difference of pressure in the air P_a and in the liquid P_l, is approximately

$$P_1 - P_a = \gamma_{lv}\left(\frac{1}{R} - \frac{1}{r}\right)$$

If $r < R$ then $P_1 - P_a$ is negative and the pressure in the air is greater than the pressure in the liquid. The two solids are held together with a pressure, $P_a - P_1$. Thus, any thin film of fluid that wets two solids will act as an adhesive. It is seen from the equation that the thinner the film, the greater the adhesion; however, formation of very thin layers requires fluids very low in viscosity. Such thin fluid films can produce high tensile strengths but are relatively weak in shear. An example is Johansson blocks that are "wrung together." These steel blocks, used as measurement standards by machinists, have exceptionally flat and smooth surfaces. If two blocks are slid or wrung together carefully, they will bond with surprising strength. It has been shown that the adsorbed water vapor, combined with the flat, smooth surface, produces a joint like the one shown, with water as the adhesive.

Estimate the force required to quickly separate two Johansson blocks that are 5 cm × 5 cm by application of a tensile load if the joint is 0.004 mm thick.

The surface energy of water is 0.080 J/m^{-2}. The density of steel is 7.8 g/cm^{-3}. Note that these blocks are square. Treat as if $R = \frac{1}{2}L$, the side of the square.

4.6 This problem deals with the solubility of silicon particles in aluminum. The solubility of large particles of silicon in aluminum at 500°C is 0.5 atom %. The solubility of aluminum in silicon can be ignored. What is the solubility in aluminum of silicon particles with a diameter of 100 Å?

The interfacial energy of silicon-aluminum is 0.42 J/m² at 500°C.

	Al	Si
Atomic mass, g/mol	27	28
Density, g/cm³	2.70	2.33

4.7 A silver wire of 1 μm diameter is suspended in a vacuum furnace at a temperature of 1000 K.

(a) What is the vapor pressure of the wire relative to a flat plate of silver at the same temperature?

(b) What is the pressure in the wire?

DATA

For silver:

Atomic weight = 107.88 g/mol
Density = 10.49 g/cm³
Surface energy at 1000 K = 1140 ergs/cm²

4.8 A silver wire 2 μm in diameter and a solid, flat silver plate are immersed in a solution of silver nitrate. What is the electrochemical potential of the wire relative to the silver plate?

DATA

For silver:

Atomic weight = 107.9 g/mol
Density = 10.5 g/cm³
Surface energy = 1000 ergs/cm²
Valence of silver in silver nitrate = +1.

4.9 Solid copper is heated to 1100 K in equilibrium with its own vapor. The surface near a grain boundary is found to be grooved with a dihedral angle of 158°. The surface energy of copper (γ_{s-v}) is 1600 ergs/cm² at 1100 K.

(a) What is the surface energy of the grain boundary at that temperature?

(b) At 1100 K, the interfacial energy of copper and a liquid metal, M, is 250 ergs/cm². What will happen at the grain boundary in part a if the solid copper is heated to 1100 K in contact with liquid metal, M?

4.10 A two-dimensional rectangular crystal has sides L_1 and L_2. The surface energies of the sides are γ_1 and γ_2, respectively. The area of the crystal, L_1L_2, is constant.

(a) Write an expression for the surface free energy of the crystal.

(b) What is the equilibrium shape of the crystal (i.e., what is the ratio of L_1 to L_2 in terms of the surface energies of the sides)?

Chapter 5

Diffusion

For microstructural changes or chemical reactions to take place in solids, there must be movement of atoms or molecules. This movement of atoms or molecules is called *diffusion*. The overall phenomenological equations that describe the motion of materials (Fick's first and second laws) are independent of the medium in which the diffusion is taking place. However, it is quite apparent that the process of diffusion takes place by different mechanisms in different media. The mechanism in gases is different from that in liquids and especially different from the diffusion mechanism in solids. In this chapter we consider first the phenomenological description of diffusion, then deal with some special aspects of diffusion of importance in solids.

5.1 FICK'S FIRST LAW

In a single-phase material, matter will flow so as to eliminate concentration gradients. Fick's first law relates the rate of flow to the concentration gradient. The relationship is stated in terms of the *flux* of a material J, the mass of the material flowing per unit time per unit area.

$$J_i = \frac{m_i}{A\tau} \tag{5.1}$$

Fick's first law states that the flux of a material, i, is linearly related to the concentration gradient of i. The constant relating the two is D_i, the diffusion coefficient. In mathematical terms, for movement in one dimension (the x direction)[1] is

$$J_{i,x} = -D_i \text{ grad } C_i = -D_i \left(\frac{\partial C_i}{\partial x} \right)_T \tag{5.2}$$

The dimensions of flux are mass (or moles) per unit area per unit time. The dimensions of the concentration gradient are mass per unit volume divided by the fourth power of length, or m/L^4. Thus the dimensions of the diffusion coefficient are length squared divided by time, typically centimeters squared per second, or meters squared per second in SI units.

$$J \left(\frac{m}{L^2 t} \right) = -D \left(\frac{L^2}{t} \right) \frac{\partial C}{\partial x} \left(\frac{m}{L^3} \frac{1}{L} \right)$$

A physical feeling for Fick's first law and the diffusion coefficient may be obtained by considering the movement of atoms between two parallel planes in a solid (Figure 5.1). There are n_1 atoms per unit area on plane 1 and n_2 atoms per unit area on plane 2. The two planes are separated by a distance, α. Assume that atoms on either plane move from that plane with a specified jump frequency Γ and that the jump frequency is the same in both directions.

The rightward and leftward movements of atoms are as follows:

$$\overrightarrow{J} = \frac{n_1 \Gamma}{2} \quad \text{and} \quad \overleftarrow{J} = \frac{n_2 \Gamma}{2}$$

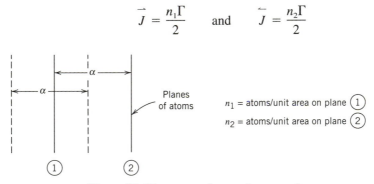

Planes of atoms

n_1 = atoms/unit area on plane (1)

n_2 = atoms/unit area on plane (2)

Figure 5.1 Movement of atoms between planes.

---

[1]Equation 5.2 is valid in all directions for an isotropic substance. It should be recognized, however, that the flux J_i is, in general, a vector. So is the gradient of C_i. The relationship between the two is expressed as a second-order tensor, which takes into account the variation of the diffusion coefficient with direction. Diffusion through graphite provides an example of such a variation. Graphite has a planar structure. Diffusion parallel to the planar direction is more rapid than diffusion perpendicular to the planes.

The net movement of atoms is

$$J = \tfrac{1}{2}(n_1 - n_2)\Gamma$$

For a unit cross-sectional area $(A = 1)$, the concentration C is related to the dimensions of the crystal α, the interatomic spacing, as follows:

$$C_1 = \frac{n_1}{\alpha} \quad \text{and} \quad n_1 = C_1\alpha$$

The flux can be stated in the following terms:

$$J_1 = \tfrac{1}{2}\alpha(C_1 - C_2)\Gamma$$

$$C_1 - C_2 = -\alpha \left(\frac{\partial C}{\partial x}\right)_T$$

$$J_1 = -\tfrac{1}{2}\alpha^2\Gamma \left(\frac{\partial C}{\partial x}\right)_T$$

Comparing this expression to Fick's first law yields:

$$D = \tfrac{1}{2}\alpha^2\Gamma \tag{5.3}$$

Typically the interplanar spacing in the solid, α, is of the order of 10^{-8} cm. For interstitial atoms such as carbon in α-iron at 900°C, D is of the order 10^{-6} cm²/s. Thus the jump frequency is about 10^{10} per second. In most metals near their melting points, diffusion coefficients are of the order of 10^{-8} cm²/s. This implies a jump frequency of about 10^8 per second.

We should differentiate between jump frequency and the vibration frequency of atoms in a solid. Vibration frequencies are on the order of 10^{13} to 10^{14} per second. Only a small portion of the vibrational movements results in a diffusional jump.

To make a successful jump, an atom must not only acquire the necessary energy, but there must also be a place into which it can settle after the movement. This is not important in the case of interstitial atoms, such as carbon in iron, because the probability that neighboring interstitial sites are occupied is low. The situation is different in the movement of substitutional atoms by a vacancy mechanism. In that case there must be a vacancy present on the site to which the diffusing atom moves. Thus the probability of a successful jump depends not only on the forces that govern the motion of the atom itself, but also on the probability that a vacancy will exist in the place to which the atom tries to move. The consequences of this point are considered later in this chapter (Section 5.7).

A simple example of the application of Fick's first law is in the analysis of the diffusion of helium through a silica wall. Let us calculate the steady state rate of diffusion of helium through a silica plate 1 mm thick at 500°C. On one side of the silica plate the helium pressure is one atmosphere. On the other side it is zero. At

500°C the solubility of helium in silica is 1.8×10^{-6} g/cm³ and the diffusion coefficient is 10^{-7} cm²/s. If the rate of dissolution on the helium side of the silica is fast enough to keep up with the diffusion through the solid, then the concentration of helium on that side is simply its solubility at a pressure of one atmosphere. The rate of helium flow is:

$$J_{He} = -D_{He} \frac{\Delta C}{\Delta x} = -10^{-7} \frac{1.8 \times 10^{-6}}{0 - 0.1} = 1.2 \times 10^{-12} \text{ g/(s·cm}^2)$$

In the case of diffusion through thin films, the steady state motion of gaseous atoms or molecules through the film is often described by a permeability constant P^*, rather than a diffusion coefficient, although the two are related. The permeability term takes solubility of the atoms or molecules into account.

Consider first the diffusion of a gas whose molecular form does not change upon dissolving in the film through which it is moving. This is the case of diffusion of gaseous molecules, such as hydrogen (H_2) through polymers, or the diffusion of monatomic gases, such as helium, through practically any material. In this case the concentration of the diffusing molecule at the surface of the film is related to the pressure of the molecule as follows:

$$H_2 = \underline{H}_2 \text{ (in solution)}$$

The equilibrium constant K, in terms of the concentration of hydrogen $[\underline{H}_2]$ and pressure of hydrogen at the surface P_{H_2}, is:

$$K = \frac{[\underline{H}_2]}{P_{H_2}} = \frac{S}{P_{H_2}}$$

where S is the solubility.

From the equation above, the relationship between solubility and pressure is simply $S = KP$, where K is the equilibrium constant.

In a film of thickness δ, with a gas pressure of the diffusing molecule P_2 on one side and P_1 on the other (Figure 5.2), the gas flux through the film is:

$$J_x = -D \frac{S_2 - S_1}{\delta}$$

$$J_x = -\frac{DK}{\delta}(P_2 - P_1)$$

If we define the permeability P^* as

$$P^* = DK \tag{5.4a}$$

then

$$J_x = -P^* \frac{P_2 - P_1}{\delta} \tag{5.4b}$$

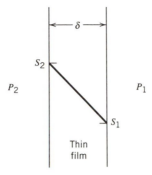

Figure 5.2 Concentration
of diffusing gas through a
thin film.

The units of permeability that one encounters in the literature can be confusing. Solubility data for gases in solids often are given as cm^3 (STP)/cm^3, that is, cubic centimeters of gas at standard temperature and pressure dissolved per cubic centimeter of solid. (STP generally means 298 K and one atmosphere pressure.) The units of the equilibrium constant K are cm^3 (STP)/$cm^3 \cdot atm$. Permeability, the product of K and the diffusion coefficient (cm^2/s), has units of cm^3 (STP)/(cm·s·atm). Additional confusion may arise if pressure is measured in units other than pascals or atmospheres—for example, centimeters of mercury. In one reference the permeability coefficient of helium in silica at 700°C is given as 2.6×10^{-12} cm^3 (STP)/(cm·s·cmHg).

It is important to note that the permeability coefficient, like the diffusion coefficient and the equilibrium constant, is a function of temperature.

Consider now the case of the diffusion of a diatomic gas, like hydrogen, through a metal film. Hydrogen, a diatomic gas, dissolves in metals in the monatomic form:

$$\tfrac{1}{2}H_2 = \underline{H} \text{ (in solution)}$$

The solubility varies with the square root of the hydrogen pressure (Sievert's law):

$$S = KP^{1/2} \tag{5.5a}$$

Accordingly, the flux of hydrogen through a metal film of thickness δ is

$$J_x = -P* \frac{P_2^{1/2} - P_1^{1/2}}{\delta} \tag{5.5b}$$

where $P*$ is often given in units of cm^3 (STP)/(cm·s·atm$^{1/2}$).

Flow rates through multilayer media can be calculated using the technique illustrated in Chapter 2 (Section 2.14, Eq. 2.43, and Appendix 2A). The derivation in

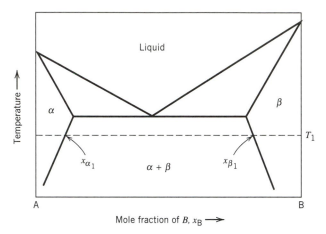

Figure 5.3 Form of phase diagram for an A–B system.

Appendix 2A, which used thermal conduction as an example, demonstrates that the total resistance to flow is simply the sum of the resistances in the individual layers. Special care must be taken in the case of diffusion of matter because the solubility of the diffusing species may be different in different layers of the diffusion medium. As an example, consider the diffusion of element B through a sample consisting of an A–B mixture. Assume that the composition of B is held at zero at one face of the sample and pure B at the other. Suppose the A–B phase diagram has the form of Figure 5.3 and that the diffusion is taking place at temperature T_1. The composition profile in the sample will be of the form shown in Figure 5.4. There will be a discontinuity at the α–β interface. The concentration gradients in the α and β phases

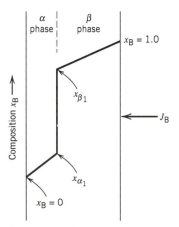

Figure 5.4 Composition profile in an A–B system when B diffuses through the sample.

are related to the diffusion coefficients of B in the phases as follows:

$$J_B = -D_{B,\beta}\left(\frac{\partial C}{\partial x}\right)_\beta = -D_{B,\alpha}\left(\frac{\partial C}{\partial x}\right)_\alpha$$

$$\frac{D_{B,\alpha}}{D_{B,\beta}} = \frac{\left(\dfrac{\partial C}{\partial x}\right)_\beta}{\left(\dfrac{\partial C}{\partial x}\right)_\alpha}$$

where $D_{B,\alpha}$ and $D_{B,\beta}$ are the diffusion coefficients of B in phases α and β.

5.2 FICK'S SECOND LAW

Fick's first law is valid on a microscopic scale; that is, it relates the mass flow rate of a component to its concentration gradient at any location. On a macroscopic scale, the first law is especially useful in the analysis of mass flow in steady state situations, where the concentration of the diffusing component changes with location but *does not change with time.* In the analysis of steady flow through a thin section of material, for example, we are interested in calculating the flow rate through the material, not in changes within the material. In many situations, however, there arises the need to calculate concentration changes with *time as well as location.* These non–steady state mass movements are the subject of Fick's second law.

Consider the motion of matter into and out of a volume V, of thickness dx, of unit cross-sectional area A (Figure 5.5). The flux entering through the left-hand side is J_1. The matter entering as J_1 either is retained in the volume or exits through the plane on the right as the flux J_2. From a mass balance, the retained mass in V is:

$$dm_i = (J_{1,i} - J_{2,i})A \, dt \qquad\qquad\qquad \textbf{(5.6)}$$

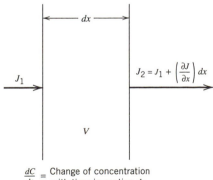

$$\frac{dC}{dt} = \text{Change of concentration with time in section } dx$$

Figure 5.5 Fick's second law.

Recognizing that the concentration is mass per unit volume, $C_i = m_i/V$, and that $V = A\,dx$:

$$\frac{dm_i}{V} = dC_i = (J_{1,i} - J_{2,i})\frac{dt}{dx}$$

The relationship between the fluxes J_2 and J_1 may also be written as follows:

$$J_{2,i} = J_{1,i} + \frac{\partial J_i}{\partial x}\,dx$$

Hence:

$$dC_i = -\frac{\partial J_i}{\partial x}\,dt \quad \text{or} \quad \frac{dC_i}{dt} = -\frac{\partial J_i}{\partial x} \tag{5.7}$$

Thus the rate of change of concentration with time at a specific location is

$$\frac{\partial C_i}{\partial t} = -\frac{\partial}{\partial x}\left(-D\,\frac{\partial C_i}{\partial x}\right) = \frac{\partial}{\partial x}\left(D\,\frac{\partial C_i}{\partial x}\right) \tag{5.8}$$

In the special case of a constant diffusion coefficient (independent of concentration), Fick's second law becomes:

$$\frac{\partial C_i}{\partial t} = D\,\frac{\partial^2 C_i}{\partial x^2} \tag{5.9}$$

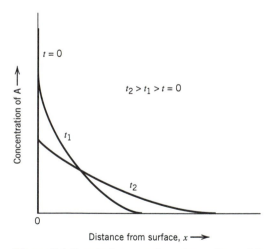

Figure 5.6 Penetration at various times of material A deposited on the surface of material B (Eq. 5.10).

This second-order partial differential equation (Eq. 5.9) may be solved for various boundary conditions. For example, if a quantity, α per unit area, of a material A is placed on one end of a semi-infinite bar of material B and then allowed to diffuse, the concentration of A as a function of time, distance from the surface (Figure 5.6), and the diffusion coefficient is given by[2]:

$$C_A(x,t) = \frac{\alpha}{\sqrt{\pi Dt}} \exp\left(-\frac{x^2}{4Dt}\right) \qquad (5.10)$$

Taking the natural logarithm of both sides of Eq. 5.10 yields:

$$\ln C_A = \ln \alpha - \tfrac{1}{2}\ln(\pi Dt) - \frac{x^2}{4Dt}$$

Thus a graph of the natural logarithm of concentration (C_A) against the square of the distance from the surface (x^2) should yield a straight line whose slope is $-1/4Dt$. The time of diffusion, t, is presumably known, and the diffusion coefficient can be calculated.

By using radioactive isotopes, this technique is often applied in the determination of the self-diffusion coefficient of a material—the diffusion coefficient of copper in copper, for example. The radioactive isotope is assumed to have the same movement characteristics as the nonradioactive material itself. Chemical analysis in this case is quite simple. It involves just the counting of radioactive disintegration rate of various sections of the semi-infinite bar after diffusion.

In the preceding example a quantity of material was deposited on the surface of

---

[2]The equation describing diffusion with this set of boundary conditions is similar to the one derived for the one-dimensional random walk (Section 2.15)

$$P(x,n) = (2\pi Nl^2)^{-1/2} \exp\left(-\frac{x^2}{2Nl^2}\right) \qquad (2.55)$$

Equation 2.55 assumes motion in both positive and negative directions. Equation 5.10 under similar conditions becomes:

$$C_A(x,t) = \frac{\alpha}{2\sqrt{\pi Dt}} \exp\left(-\frac{x^2}{4Dt}\right)$$

Comparing the two terms within the exponential term yields:

$$D = \frac{1}{2}\frac{N}{t}l^2$$

where N/t is the jump frequency in the x direction (denoted as Γ in Section 5.1, Eq. 5.3), and l is the jump distance (denoted as α in Eq. 5.3).

a solid and then allowed to diffuse into it. The surface concentration of the diffusing material decreases with time (Figure 5.6). Fick's second law can also be used to analyze the diffusion of one material into another when the concentration of the diffusing material, A, is held constant at the surface of the material, B, through which it is diffusing. Carburization or nitriding of steel, and the introduction of dopants to silicon, are examples of such processes. As one boundary condition, assume that the sample into which material A is being added is semi-infinite in extent; that is, it is long compared to diffusion distances. The initial concentration of material A throughout the semi-infinite bar is taken to be C_0. The concentration of A at the surface is raised to C_s and maintained at that level for time t, greater than zero. The partial differential equation of the second law (Eq. 5.9) can be reduced to an ordinary differential equation by making the following substitution:

$$\eta = \frac{x}{2\sqrt{Dt}}$$

Equation 5.9 then becomes an ordinary differential equation in c and η:

$$\frac{dC}{d\eta} = -\frac{1}{2\eta}\frac{d^2C}{d\eta^2}$$

The solution to this equation with the stated boundary conditions is:

$$\frac{C - C_0}{C_s - C_0} = 1 - \mathrm{erf}\left(\frac{x}{2\sqrt{Dt}}\right) = \mathrm{erfc}\left(\frac{x}{2\sqrt{Dt}}\right) \qquad \textbf{(5.11a)}$$

where x is the distance from the free surface, and $\mathrm{erf}(x/2\sqrt{Dt})$ is the error function of $x/2\sqrt{Dt}$, and $\mathrm{erfc}(x/2\sqrt{Dt})$ is the complementary error function of $x/2\sqrt{Dt}$. The properties of the error function and some representative values are given in Appendix 5A. A schematic plot of Eq. 5.11a is shown in Figure 5.7.

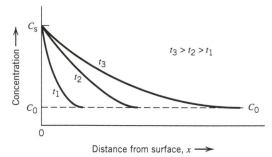

Figure 5.7 Concentration profile after diffusion for time t with surface concentration maintained at C_s (Eq. 5.11a).

The relation between diffusion distance and time in Eq. 5.11a is especially important from a practical point of view. Let us define the depth of penetration of A into B as the location (x) where the concentration of A becomes C_A^*. If $C = C_A^*$, the concentration term on the left-hand side of Eq. 5.11a is fixed. The term $x/2\sqrt{Dt}$ is then also fixed. That means that the penetration distance is a function of the *square root* of the diffusion time. For example, if a diffusion penetration of 0.1 mm develops in one hour, it will take 4 hours to develop a penetration of 0.2 mm.

As a special case of Eq. 5.11a consider a semi-infinite solid containing no material A at the outset ($C_0 = 0$ at $t = 0$). If the concentration of A on the free surface C_s is held constant after time $= 0$, the concentration of A as a function of position and time is given by

$$C(x,t) = C_s \left[1 - \text{erf}\left(\frac{x}{2\sqrt{Dt}}\right) \right] = C_s \, \text{erfc}\left(\frac{x}{2\sqrt{Dt}}\right) \qquad \textbf{(5.11b)}$$

The total amount of A that enters during time t, Q_A, is simply the integral of Eq. 5.11b over all x.

$$Q_A = \int_0^\infty C_s \, \text{erfc}\left(\frac{x}{2\sqrt{Dt}}\right) dx$$

An equivalent, and mathematically simpler, way to evaluate Q_A is to integrate J_A, the flux of A, through the surface plane ($x = 0$) for the total diffusion time t. By differentiating Eq. 5.11b and evaluating at $x = 0$, we have

$$J_{x=0} = C_s \left(\frac{D}{\pi t}\right)^{1/2}$$

$$Q_A = \int_0^t J_{x=0} dt = \int_0^t C_s \left(\frac{D}{\pi t}\right)^{1/2} dt = 2C_s \left(\frac{Dt}{\pi}\right)^{1/2} \qquad \textbf{(5.12)}$$

If two semi-infinite solids, A and B, are joined (welded) together and heated until interdiffusion takes place (Figure 5.8), it can be shown that the interface between the two will assume a composition of $C_A/2$ and will remain at that level throughout the diffusion process. The concentration of A as a function of distance is:

$$\frac{C_A}{C_{A,0}} = \frac{1}{2} \, \text{erfc}\left(\frac{x}{2\sqrt{Dt}}\right) \qquad \textbf{(5.13)}$$

It is important to emphasize that the solutions in Eqs. 5.10, 5.11a, 5.11b, 5.12, and 5.13 assume a single, *constant* diffusion coefficient for the system, called the interdiffusion coefficient, D. This interdiffusion coefficient, also written as $\tilde{D}$, need not be constant as a function of composition. We show next that D is, in fact, not constant in many solid state diffusion cases.

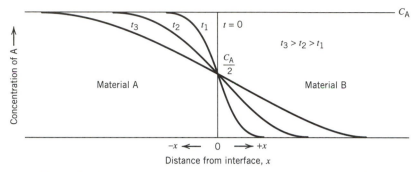

Figure 5.8 Interdiffusion of A and B with constant diffusion coefficient.

5.3 KIRKENDALL EFFECT

In the foregoing case of interdiffusion it was assumed that the movements could be described by a single diffusion coefficient. In terms of atom movements in a two-component (A–B) solution, this means that as atoms of A move in one direction, an equivalent number of B atoms move in the opposite direction. This condition is true in ideal gases and between liquids with equal molar volumes, but it may or may not be true in solids.

One of the most significant experiments in the field of solid state diffusion was conducted in 1947 when Smigelskas and Kirkendall (Ref. 1) demonstrated that two materials diffusing into each other need not have the same diffusion coefficient. That is, if two materials, A and B, are joined, the diffusion coefficient of A is not necessarily equal to the diffusion coefficient of B. This inequality of diffusion rates should be intuitively plausible in cases such as interstitial diffusion. For example, carbon dissolves interstitially in iron. If a block of carbon were attached to a block of iron, and the two held at a high temperature to allow diffusion to take place, carbon would move rapidly through the iron but the iron would move hardly at all through the carbon. The two diffusion movements are independent because the carbon atoms and iron atoms occupy different lattice positions, carbon atoms being found in interstitial positions in the iron lattice. (Note that when this type of diffusion takes place, the interface between carbon and iron moves toward the carbon.)

What is particularly striking about the Kirkendall experiment is that it demonstrated that the same relationship may hold true for materials forming *substitutional* solutions, in which the atoms A and B share the same lattice. For this to be true—that is, to have the diffusion coefficient of one greater than the other—requires that vacancies be present and that local sources and sinks for vacancies be available. Indeed Kirkendall's experiments were crucial in demonstrating that vacancies exist in metals.

To get a physical feeling for the consequences of different diffusion rates, consider a hypothetical experiment of the movement of two gases through a membrane that is supported in a tube and is free to slide left or right (Figure 5.9). Suppose that the diffusion of gas B on the right is faster than the diffusion rate of gas A through the membrane. The flux of B to the left will be greater than the flux of A to the right. This will increase the pressure on the left-hand side, and the membrane, which is free to slide, will move to the right. Thus the membrane movement will be a measure of the relative motion of the gas molecules through the membrane.

The same relationship holds in solids. If two materials are welded together and markers placed at the interface between them, the markers will move if the diffusion rates of the two materials are different. This was the basis of Kirkendall's experiment.

To analyze this situation (Ref. 2), first take the viewpoint of someone sitting on the marker. The fluxes of atoms A and B will be[3]:

$$J_A = -D_A \frac{dC_A}{dx} \quad \text{and} \quad J_B = -D_B \frac{dC_B}{dx}$$

The sum of the fluxes yields

$$J_A + J_B = \frac{\Delta m_A + \Delta m_B}{A\Delta t} = \frac{\Delta m}{A\Delta t} \tag{5.14a}$$

The concentration C is mass (m) divided by volume (V), hence:

$$C = \frac{m}{V} \quad m = CV \quad \Delta m = C\Delta V$$

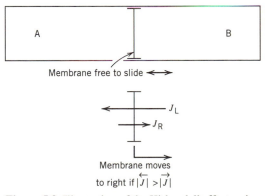

Figure 5.9 Illustration of the Kirkendall effect, using the diffusion of gases through a membrane.

[3]In these equations and those that follow we will use the "d" notation for differentials. In Section 5.4, when partial derivative notation "∂" is used, the condition of constant temperature applies.

The net mass passing through the interface is Δm with a volume ΔV. This ΔV is $-A\Delta l$. Note that Δl divided by time is the velocity of the interface or the markers at the interface. The concentration at the interface remains constant, experimentally.

$$\frac{\Delta m}{A\Delta t} = \frac{C\Delta V}{A\Delta t} = -\frac{CA\Delta l}{A\Delta t} = -C\frac{\Delta l}{\Delta t}$$

$$(5.14b)$$

$$\frac{\Delta m}{A\Delta t} = -Cv$$

where v is the velocity of the marker. The marker moves in a direction opposite to the net mass flow.

Thus from Eqs. 5.14a and 5.14b:

$$-Cv = J_A + J_B = -D_A\frac{dC_A}{dx} - D_B\frac{dC_B}{dx}$$

But $C_A + C_B = C$, the total concentration of atoms (total moles per unit volume) is constant, therefore:

$$\frac{dC_A}{dx} + \frac{dC_B}{dx} = 0$$

$$(5.15)$$

$$v = \frac{1}{C}(D_A - D_B)\frac{dC_A}{dx}$$

$$(5.16)$$

Now consider the movements of atoms relative to the end of the bar of material (bar is long relative to diffusion distances), instead of relative to the markers. Fick's first law is now written as follows:

$$J_A = -D_A\frac{dC_A}{dx} + vC_A$$

$$(5.17)$$

where v is the velocity of the marker.

From Fick's second law:

$$\frac{dC_A}{dt} = -\frac{dJ_A}{dx} = -\frac{d}{dx}\left[-D_A\frac{dC_A}{dx} + vC_A\right]$$

$$(5.18)$$

Combining with Eq. 5.16, we have

$$\frac{dC_A}{dt} = -\frac{d}{dx}\left[-D_A\frac{dC_A}{dx} + C_A\left(\frac{1}{C}[D_A - D_B]\frac{dC_A}{dx}\right)\right]$$

$$(5.19)$$

With some mathematical rearrangements:

$$\frac{dC_A}{dt} = \frac{d}{dx}\left[\left(\frac{C_B}{C} D_A + \frac{C_A}{C} D_B \right) \frac{dC_A}{dx} \right]$$

$$\frac{dC_A}{dt} = \frac{d}{dx}\left[(N_B D_A + N_A D_B) \frac{dC_A}{dx} \right]$$

(5.20a)

where $N_A = C_A/C$.

Comparing Eq. 5.20a to the expression for Fick's second law (Eq. 5.8), the interdiffusion coefficient, $\tilde{D}$, is

$$\tilde{D} = N_B D_A + N_A D_B$$

(5.20b)

This is equivalent to saying that in an A–B diffusion couple, if the flux of B, J_B, is not equal to the flux of A, J_A, there exists some moving reference frame in which the fluxes of the two are equal. In that reference frame, the diffusion coefficients of A and B are equal to each other and equal to the interdiffusion coefficient, $\tilde{D}$.

In his original experiment, Kirkendall observed the diffusion of copper and zinc in a diffusion couple consisting of copper and brass, a copper–zinc solid solution alloy. He marked the interface with fine molybdenum wires and observed their movement after the diffusion had taken place.[4] The markers moved toward the brass side, leading Kirkendall to conclude that the zinc moved faster than the copper, in fact, three times as fast.

Of course the existence of different diffusion rates for A and B requires that vacancies be present in the materials if A and B form a set of substitutional alloys. Because B is diffusing faster into A than A into B, vacancies will have to be created on the A side and destroyed on the B side for an equilibrium number of vacancies to be present on each side. It may not be possible in an actual experiment to create and destroy vacancies at the rate required. In fact, in Kirkendall's experiment, where zinc diffused roughly three times as fast as copper, actual voids were observed on the high-zinc side. Kirkendall concluded that the brass was unable to destroy the vacancies fast enough to prevent the formation of physical voids, which were observed in a microscopic section of the diffusion samples following the experiment. This effect is sometimes called Kirkendall porosity.

5.4 MOBILITY

In the preceding sections, it was assumed that diffusion rates are proportional to concentration gradients. This analysis is valid for ideal solutions but must be modified for nonideal solutions. To illustrate, consider the phase separation that occurs in a miscibility gap illustrated in Figure 5.10. At temperature T_1 a material with an

---

[4]Molybdenum does not dissolve in the copper or the brass.

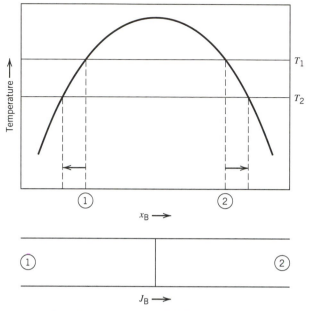

Figure 5.10 Illustration of "uphill" diffusion: material B will tend to move from point 1 to point 2 upon cooling from temperature T_1 to T_2.

overall composition in the miscibility gap will, at equilibrium, consist of two phases. The phase compositions are labeled 1 and 2 in Figure 5.10. As the temperature is lowered to T_2, the equilibrium phase compositions will change. The material labeled 2 will increase in its concentration of B. In fact the material will diffuse from material 1 (low B concentration) into the material 2, of high B concentration (i.e., "uphill" diffusion). Consider Fick's first law:

$$J_B = -D_B \frac{dC_B}{dx} \tag{5.2}$$

The flux J_B is positive. The concentration gradient for B is also positive. For the equation to apply, therefore, the diffusion coefficient would have to be negative, which doesn't make much sense. One can conclude that the driving force for diffusion cannot be simply the concentration gradient.

It was first suggested by Einstein that diffusion is driven by a virtual force, which is the negative gradient of the chemical potential (or partial molar free energy).

$$F_i = -\frac{1}{N_A} \left(\frac{\partial \mu_i}{\partial x} \right)_T \tag{5.21}$$

where N_A is Avogadro's number, and F_i is the virtual force on the diffusing species.

The flux of the diffusing species i is given by

$$J_i = v_i C_i \tag{5.22}$$

where for the species i, v_i is the drift velocity, C_i is the concentration, and J_i is the flux.[5]

The absolute mobility of the diffusing species, B_i, is defined as its velocity divided by the force

$$B_i \equiv \frac{v_i}{F_i} \tag{5.23}$$

Combining Eqs. 5.21, 5.22, and 5.23 yields:

$$B_i = -\frac{v_i}{\frac{1}{N_A}\left(\frac{\partial \mu_i}{\partial x}\right)_T}$$

and

$$J_i = -\frac{B_i}{N_A}\left(\frac{\partial \mu_i}{\partial x}\right)_T C_i \tag{5.24}$$

Comparing Eq. 5.24 to the statement of Fick's first law (Eq. 5.2) yields, at constant temperature:

$$-D_i \frac{dC_i}{dx} = -\frac{B_i}{N_A}\frac{d\mu_i}{dx} C_i$$

$$D_i = \frac{B_i}{N_A}\frac{d\mu_i}{dC_i} C_i \tag{5.25}$$

Remembering that the chemical potential is:

$$\mu_i = \mu_i^\circ + RT \ln a_i = \mu_i^\circ + RT \ln(\gamma_i C_i)$$

---

[5]Confusion sometimes arises if one thinks of the relationship between force and acceleration from the laws of motion of free bodies in which $F = Ma$. In the case considered above, there is no acceleration. It is assumed that the velocity is a constant *drift* velocity. The particle is assumed to have a resisting force that is proportional to its velocity. The particle travels at the net velocity v_i, which just balances the virtual force created by the negative gradient of the chemical potential.

then:

$$D_i = \frac{B_i RT}{N_A} \frac{d(\ln a_i)}{d(\ln C_i)}$$

$$D_i = B_i kT \left[\frac{d(\ln C_i) + d(\ln \gamma_i)}{d(\ln C_i)} \right]$$
(5.26)

From this we obtain the final relationship between the diffusivity of a species, its absolute mobility, and its thermodynamic activity coefficient.

$$D_i = B_i kT \left[1 + \frac{d(\ln \gamma_i)}{d(\ln C_i)} \right]$$
(5.27)

In the case of ideal solutions, or in dilute solutions where the activity coefficient is constant (Henry's law), the diffusivity is simply

$$D_i = B_i kT$$
(5.28)

The self-diffusion coefficient of a material, which is a measure of the absolute mobility of the diffusing species, may be determined using radioactive tracers. It is assumed that the various isotopes of an element diffuse at the same rate. The movement of radioactive isotopes is easy to follow because the particles can be detected through the darkening of photographic film, or by relatively simple disintegration counting techniques. If we let the asterisk superscript represent the quantities pertinent to the radioactive tracer, we can write Eq. 5.27 as follows:

$$D_i^* = B_i^* kT \left[1 + \frac{d(\ln \gamma_i^*)}{d(\ln C_i)} \right]_{N_i + N_i^*}$$
(5.29)

Note that γ, the thermodynamic activity coefficient, is a function of the total concentration of i atoms, the sum of radioactive and nonradioactive i. It is not a function of the very dilute concentration of radioactive i, N_i^*. Thus:

$$D_i^* = B_i^* kt$$
(5.30a)

Combining with Eq. 5.27 yields

$$D_i = D_i^* \left[1 + \frac{d(\ln \gamma_i)}{d(\ln C_i)} \right]$$
(5.30b)

Combining all of this with the equation from the Kirkendall experiment (Eq. 5.20a) yields

$$\tilde{D} = N_B D_A^* \left[1 + \frac{d(\ln \gamma_A)}{d(\ln C_A)} \right] + N_A D_B^* \left[1 + \frac{d(\ln \gamma_B)}{d(\ln C_B)} \right]$$

From the Gibbs–Duhem equation:

$$N_A d\mu_A + N_B d\mu_B = 0$$

Note that $dN_A = -dN_B$

$$N_A \frac{d(\ln \gamma_A)}{dN_A} = N_B \frac{d(\ln \gamma_B)}{dN_A}$$

Thus:

$$\tilde{D} = (N_B D_A^* + N_A D_B^*) \left[1 + \frac{d(\ln \gamma_A)}{d(\ln C_A)} \right] \tag{5.31}$$

The interdiffusion coefficient, $\tilde{D}$, is thus a function of composition in a binary system, as well as the thermodynamic characteristics of the solutions formed between the two components. The analysis of diffusion concentration profiles under these conditions requires more advanced techniques.

5.5   DIFFUSION IN IONIC SOLIDS

The concept of mobility is especially useful in dealing with conductivity in ionic solids. If we define

C_i as the concentration of charged particles per unit volume
v_i as the drift velocity
Z_i as the valence of the particle ($Z_i e$ = charge of particle)

then $J_{q,c}$, the drift velocity of electrical charge related to the motion of the i species is:

$$J_{q,i} = v_i C_i Z_i e \tag{5.32}$$

Conductivity is defined as the ratio of mobility to the electric field, that is:

$$\sigma_i = \frac{J_{q,i}}{E} = C_i Z_i e \frac{v_i}{E} \tag{5.33}$$

Mobility is defined as:

$$B_i = \frac{v_i}{F_i} \tag{5.34}$$

and the force, F_i, on the species is $Z_i eE$ (neglecting any chemical driving force), or

$$B_i = \frac{v}{Z_i eE} \qquad (5.35)$$

where Z_i is the valence of i.

Therefore, in terms of absolute mobility, conductivity is

$$\sigma_i = C_i Z_i^2 e^2 B_i \qquad (5.36)$$

When multiple species are involved in the transport of electrical charge, the fraction of charge carried by the species i is t_i, the transference number of i.

$$\sigma_i = \sigma t_i \qquad (5.37)$$

The diffusion coefficient in terms of mobilities is:

$$D_i = B_i kT \qquad (5.28)$$

Hence:

$$D_i = \frac{\sigma t_i kT}{C_i Z_i^2 e^2} \qquad (5.38)$$

Equation 5.38 gives the relation of the diffusion coefficient to the electrical conductivity contribution of the i species.

5.6   TEMPERATURE DEPENDENCE OF DIFFUSION

Empirically it is found that the relationship between D, the diffusion coefficient, and temperature can usually be described by the equation:

$$D = D_0 \exp\left(-\frac{Q}{RT}\right) \qquad (5.39)$$

In this equation, D_0 and Q may vary with composition of the materials involved but are generally independent of temperature. The values of these two variables can be obtained by plotting the natural logarithm of D versus $1/T$. The slope of this plot yields a value of Q/R.

$$\ln D = \ln D_0 - \frac{Q}{R}\left(\frac{1}{T}\right) \qquad (5.40)$$

The extrapolation of the line to $1/T = 0$ yields the natural logarithm of D_0.

It is apparent that the form of Eq. 5.40 is similar to that involving reaction rate constants (Arrhenius relationship). The factor Q is the activation energy required for diffusion. One can think of it as a quantity of energy required by the diffusing species to overcome the energy barrier separating two atomic planes. In the case of interstitial diffusion, this factor Q can simply be equated to the energy required for motion, ΔH_M. It is assumed that interstitial solutions are rather dilute and that the atom diffusing via an interstitial mechanism will not encounter another interstitial in the position to which it is moving.

The same is not true of diffusion involving a vacancy mechanism. For an atom to move into a new position in that case, two conditions must be satisfied. First, it must overcome the activation energy barrier, and second, it must encounter a vacancy in the direction of desired movement. Based on statistical thermodynamics, the probability of finding a vacancy on a particular lattice site is proportional to

$$P_v = \exp\left(\frac{-\varepsilon_v}{kT}\right) \quad \text{or} \quad \exp\left(\frac{-\Delta H_v}{RT}\right)$$

The diffusion coefficient as it relates to temperature is then

$$D = D_0 \exp\left(-\frac{\Delta H_M}{RT} - \frac{\Delta H_v}{RT}\right) \tag{5.41}$$

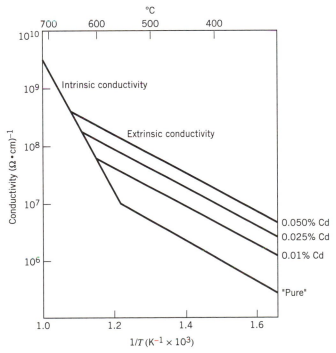

Figure 5.11 Log conductivity versus reciprocal temperature for NaCl crystals doped with varying amounts (atom %) of CdCl$_2$.

Thus the activation energy for movement via a vacancy mechanism is the sum of the enthalpies of formation of the vacancy and the activation energy for atom movement.

In ionic crystals, the factor Q is related to the activation energy for movement and the activation energy of formation of a vacancy. In the case of Schottky defects (in the intrinsic region) this value Q is equal to

$$\frac{Q}{R} = \frac{\Delta H^+}{R} + \frac{\Delta H_s}{2R} \tag{5.42}$$

In ionic crystals that have impurities at a concentration high enough to induce extrinsic conductivity, the activation energy for diffusion becomes simply $\Delta H^+/R$. This change in Q/R can be observed by measuring the diffusion coefficient, or electrical conductivity, as a function of temperature (see Figure 5.11).

5.7   HIGH DIFFUSIVITY PATHS

The foregoing analyses were all concerned with atomic movements through the bulk of a material (i.e., volume diffusion). The only structural imperfections considered were vacancies. In addition to vacancies, several types of structural imperfection are present in most materials, such as dislocations, grain boundaries, and external surfaces. Because the movement of atoms (diffusion) along these imperfections is usually more rapid than diffusion through a single crystal of a material, the imperfections are thought of as short circuits to the process, or high diffusivity paths.

One would intuitively believe that the movement of atoms along an edge dislocation is more easily accomplished than movement through a crystal containing only vacancies because the disturbance at the dislocation edge creates a more continuous open structure. If we think of a grain boundary as a collection of dislocations (at least at low-angle grain boundaries), then diffusion along grain boundaries should be more rapid than bulk diffusion. An early experiment (Ref. 3) on self-diffusion in silver demonstrated that there is indeed a difference in measured diffusivity between single crystals and polycrystalline specimens (Figure 5.12). Grain boundary diffusion (D_{poly} in Figure 5.12), which was characterized by a lower activation energy, became dominant at temperatures below 700°C. The grain boundary diffusion coefficient is larger than D_{single} the lattice diffusion coefficient, throughout the temperature range, as illustrated in Figure 5.13. The measured coefficient, of course, takes into account the total grain boundary area in the sample. Another sample, with a different grain size, would show a different grain boundary contribution to diffusivity. The quantitative analysis of diffusion through bulk and grain boundaries combined is treated in more advanced texts and publications (Ref. 4).

In addition to movement along grain boundaries, a material may diffuse along a free surface. This path is even more unimpeded than a grain boundary and can be expected to require lower activation energy than either bulk diffusion or grain boundary diffusion. The relationship among the three is shown in Figure 5.13. It

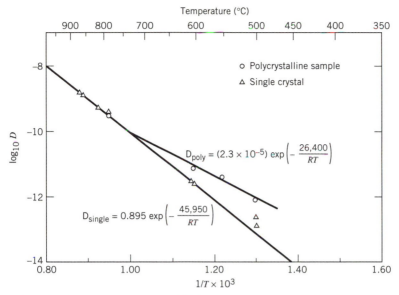

Figure 5.12 Values of the self-diffusion coefficient obtained for silver using single-crystal and polycrystalline samples. (From Ref. 3, p. 129.)

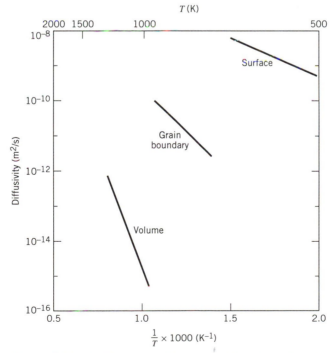

Figure 5.13 Self-diffusion coefficients for silver depend on the diffusion path. In general, diffusivity is greater through less restrictive structural regions.

is important to note that the information displayed in Figure 5.13 relates to the diffusivity along various paths. The relative total movements in a particular sample of material depend not only on the diffusivity of the paths, but also on the number or concentration of the paths. In a large-grained sample, grain boundary diffusion will be less important than in fine-grained samples because in the large-grained sample there is less grain boundary area along which to move. Similarly, surface diffusion is relatively unimportant in large samples of material. It is, however, a more important component of total diffusion in the case of thin samples, such as the ones found in thin films used in electronic applications.

A simple model (Figure 5.14) can be used to illustrate the point just made concerning the relative importance of diffusion through the lattice and through grain boundaries. The structure through which the material flows consists of square grains of side L, and grain boundaries of thickness δ. The diffusing material flows through the thickness d.

Lattice diffusion is characterized by

$$J_{\mathrm{L}} = \frac{m_{\mathrm{L}}}{L^2 t} = -D_{\mathrm{L}} \frac{dC}{dx} \quad \text{or} \quad \dot{m}_{\mathrm{L}} = -D_{\mathrm{L}} L^2 \frac{dC}{dx} \qquad (5.43)$$

where $\dot{m}$ is the rate of mass flow through the lattice, m/t.

For the grain boundaries, the relationship is[6]:

$$\dot{m}_{\mathrm{b}} = -2D_{\mathrm{gb}} \delta L \frac{dC}{dx} \qquad (5.44)$$

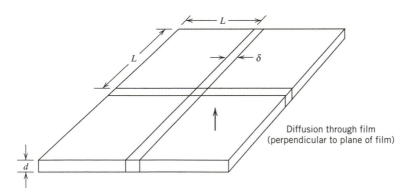

Figure 5.14 Model for diffusion through a film containing grain boundaries.

---

[6]To fill the two-dimensional space, there should be two grain boundary segments, δL, for each square grain. We will assume that the dimension δ is adjusted to account for this.

A correlation for lattice diffusion coefficients in face-centered cubic metals (Ref. 5) is

$$D_L = 0.5 \exp\left(-17 \frac{T_m}{T}\right) \text{ cm}^2/\text{s} \qquad (5.45)$$

where T_m is the melting temperature.

A similar correlation for grain boundary diffusion in face-centered cubic metals is

$$\delta D_b = 1.5 \times 10^{-8} \exp\left(-8.9 \frac{T_m}{T}\right) \text{ cm}^3/\text{s} \qquad (5.46)$$

Based on Eqs. 5.45 and 5.46, the movement of material through the grain boundaries relative to the lattice is[7]:

$$\frac{\dot{m}_{gb}}{\dot{m}_L} = \frac{2\delta D_{gb} L \left(\frac{dC}{dx}\right)_{gb}}{D_L L^2 \left(\frac{dC}{dx}\right)_L} = \left(\frac{\delta D_{gb}}{D_L}\right)\frac{1}{L}$$

$$\frac{\dot{m}_{gb}}{\dot{m}_L} = \frac{2}{L}\frac{1.5 \times 10^{-8} \exp(-8.9 T_m/T)}{0.5 \exp(-17 T_m/T)}$$

$$\frac{\dot{m}_{gb}}{\dot{m}_L} = \frac{2}{L}(3 \times 10^{-8}) \exp\left(8.1 \frac{T_m}{T}\right) \qquad (5.47)$$

From Eq. 5.47, it is apparent that smaller grain size (L), and lower temperatures favor grain boundary diffusion relative to lattice diffusion.

At a temperature of six-tenths of the melting point ($T/T_m = 0.6$), we can solve for the grain size at which the rate of flow through grain boundaries and lattice are equal.

$$\frac{\dot{m}_{gb}}{\dot{m}_L} = \frac{2}{L}(3.0 \times 10^{-8}) \exp\left(8.1 \times \frac{1}{0.6}\right) = \frac{2.2 \times 10^{-2}}{L} \text{ cm}$$

$$L = 1.1 \times 10^{-2} \text{ cm} = 110 \ \mu\text{m}$$

At the melting point (or just below), the grain size for equal flow rates through grain boundaries and lattice is about one micrometer, which is a very fine grain size.

---

[7]The concentration gradients through the lattice and the grain boundaries are equal because the concentrations on each side of the thickness are equal. We are calculating the relative rates of flow under the same concentration gradient.

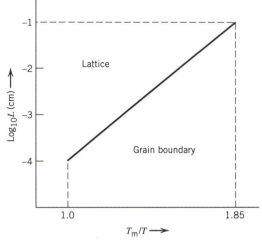

Figure 5.15 Regions of lattice-dominated and grain-boundary-dominated diffusion for the model in Figure 5.14.

At that grain size, grain boundary diffusion will dominate for all temperatures at which the material is solid. Based on this admittedly simple model, the relationship between temperature and dominant diffusion mode is shown in Figure 5.15.

5.8 FORMATION OF COMPOUNDS

Two materials, A and B, form a compound, AB, and, except for that, are essentially insoluble in each other. If a diffusion couple is made by joining pure A and pure B, the compound AB will appear between them after a period of time at a temperature that allows atoms to move (Figure 5.16).

Let us apply the concept of mobility from Section 5.4: that the flux is proportional to the negative of the chemical potential gradient for material A (Figure 5.17).

$$J_A = -L_A \text{ grad } \mu_A = -L_A \frac{\Delta\mu_A}{x} \tag{5.48}$$

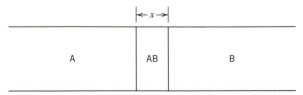

Figure 5.16 Compound formation between regions A and B.

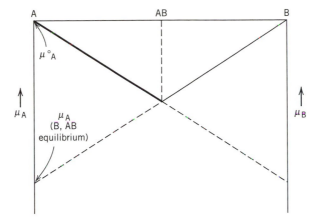

Figure 5.17 Chemical potential (μ) of A and B with compound AB.

The flux J_A is $dm_A/A\,dt$.
In the compound AB,

$$V_{AB} = m_A \overline{V}_A$$

$$dV_{AB} = \overline{V}_A\,dm_A = A\,dx$$

$$dm_A = \frac{A}{\overline{V}_A}\,dx$$

Substituting in Eq. 5.48,

$$J_A = \frac{1}{\overline{V}_A}\frac{dx}{dt} = -L_A\frac{\Delta\mu_A}{x}$$

From Figure 5.17, $\Delta\mu_A = \mu_A - \mu_A^\circ$, hence

$$x\,dx = L_A(\mu_A^\circ - \mu_A)\overline{V}_A\,dt$$

Integrating:

$$\int_0^x x\,dx = \int_0^t L_A(\mu_A^\circ - \mu_A)\overline{V}_A\,dt \tag{5.49}$$

$$x^2 = L_A(\mu_A^\circ - \mu_A)\overline{V}_A t = kt$$

where $k = L_A(\mu_A^\circ - \mu_A)\overline{V}_A$, a constant.
We can conclude from Eq. 5.49 that the thickness of the compound formed is a function of the square root of time.

5.9   BOLTZMANN–MATANO ANALYSIS

The equation of Fick's second law was integrated in Section 5.2 for several cases, all of which assumed a constant diffusion coefficient, $\tilde{D}$. In Section 5.3, the Kirkendall effect was discussed. As part of the analysis, Eq. 5.20b was derived.

$$\tilde{D} = N_B D_A + N_A D_B \tag{5.20b}$$

Based on this equation, the interdiffusion coefficient $\tilde{D}$ is not constant if the diffusivities of the two components, D_A and D_B, are unequal, which is usually the case. A technique based on the Boltzmann–Matano analysis may be used to determine the interdiffusion coefficient (or coefficients) from experimental data when $\tilde{D}$ varies with composition.

The experimental situation is shown in Figure 5.18. Two samples have been joined and then allowed to diffuse into each other. We are dealing with a two-component system, so we may represent the composition by one variable, C, which is the concentration of one of the components. The specimen is assumed to be very long compared to the diffusion zone. After the diffusion has taken place, the specimen is analyzed chemically, and the curve for composition versus distance (x) is shown in Figure 5.18. The first problem is to find the location of an interface to use as the basis for determining $\tilde{D}$, or the several values of $\tilde{D}$ as a function of composition.

The second law is[8]:

$$\frac{\partial C}{\partial t} = \frac{\partial}{\partial x}\left(\tilde{D}\,\frac{\partial C}{\partial x}\right) \tag{5.9}$$

We would like to convert the equation into one involving two variables instead of three: C, x, and t. To do so, let

$$\eta = \frac{x}{t^{1/2}} \tag{5.50}$$

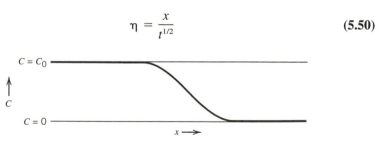

Figure 5.18 Concentration C versus distance x after diffusion.

---

[8]The constant temperature symbol is omitted from the partial derivatives because it is understood that the diffusion took place at constant temperature.

Then:

$$\frac{\partial C}{\partial t} = \frac{dC}{d\eta}\frac{\partial \eta}{\partial t} = -\frac{1}{2}\frac{x}{t^{3/2}}\frac{dC}{d\eta} = -\frac{1}{2}\frac{\eta}{t}\frac{dC}{d\eta} \tag{5.51}$$

$$\frac{\partial C}{\partial x} = \frac{dC}{d\eta}\frac{\partial \eta}{\partial x} = \frac{1}{t^{1/2}}\frac{dC}{d\eta} \tag{5.52}$$

Substituting Eqs. 5.51 and 5.52 in Eq. 5.9, we have

$$-\frac{1}{2}\frac{\eta}{t}\frac{dC}{d\eta} = \frac{\partial}{\partial x}\left(\tilde{D}\frac{1}{t^{1/2}}\frac{dC}{d\eta}\right)$$

$$\frac{\partial}{\partial x} = \frac{d}{d\eta}\left(\frac{\partial \eta}{\partial x}\right) = \frac{1}{t^{1/2}}\frac{d}{d\eta}$$

$$-\frac{1}{2}\frac{\eta}{t}\frac{dC}{d\eta} = \frac{1}{t}\frac{d}{d\eta}\left(\tilde{D}\frac{dC}{d\eta}\right)$$

$$-\frac{1}{2}\eta\, dC = d\left(\tilde{D}\frac{dC}{d\eta}\right) \tag{5.53}$$

We have reduced the second law equation to one involving two variables, η and C. Let us now integrate Eq. 5.53 from the end of the bar, where $C = 0$, to some point where $C = C'$.

$$-\frac{1}{2}\int_0^{C'} \eta\, dC = \int_0^{C'} d\left(\tilde{D}\frac{dC}{d\eta}\right) = \left[\tilde{D}\frac{dC}{d\eta}\right]_0^{C'} \tag{5.54}$$

Note that at $C = 0$, the slope of the curve, dC/dx, is zero. Because $\eta = x/t^{1/2}$, the value of $dC/d\eta$ is also zero. The same is true at the other end of the bar, where $C = C_0$. Then:

$$-\frac{1}{2}\int_0^{C_0} \eta\, dC = 0$$

$$-\frac{1}{2}\frac{1}{t^{1/2}}\int_0^{C_0} x\, dC = 0 \tag{5.55}$$

Equation 5.55 defines x; that is, it locates where $x = 0$, the interface between the two samples. This is shown in Figure 5.19. At $x = 0$ the areas on either side of $x = 0$ (one considered positive and the other negative) are equal, thus $\int_0^{C_0} x\ dC = 0$.

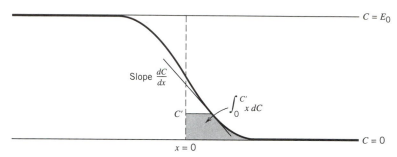

Figure 5.19 Determination of quantities for Eqs. 5.55 and 5.56.

Once having established the location of the point where $x = 0$, we can proceed to evaluate the diffusion coefficient(s). Substituting $\eta = x/t^{1/2}$ in Eq. 5.54, we have

$$-\frac{1}{2t^{1/2}} \int_0^{C'} x\, dC = \tilde{D}\, \frac{dC}{dx}\frac{dx}{d\eta} = t^{1/2}\tilde{D}\,\frac{dC}{dx}$$

$$-\frac{1}{2t} \int_0^{C'} x\, dC = \tilde{D}\,\frac{dC}{dx}$$

$$\tilde{D} = -\frac{1}{2t}\left(\frac{dx}{dC}\right)\int_0^{C'} x\, dC \qquad (5.56)$$

At a composition C', the integral $\int_0^{C'} x\, dC$ is the shaded area shown in Figure 5.19. The slope of the curve at C', dC/dx, can be determined. Presumably, the diffusion time, t, is known. Hence D, the interdiffusion coefficient, can be calculated.

REFERENCES

1. Smigelkas, A., and Kirkendall, E., *Trans. AIME, 171,* 130 (1947).
2. Darken, L., *Trans. AIME, 174,* 184 (1948).
3. Turnbull, D., *Atom Movement,* American Society for Metals, Metals Park, OH, 1951.
4. Shewmon, P. G., *Diffusion in Solids,* McGraw-Hill, New York, 1963.
5. Balluffi, R. W., and Blakely, J. M., *Thin Solid Films, 25,* 363 (1975).

APPENDIX 5A

The Error Function

The error function of a variable z, erf(z), is defined as follows:

$$\text{erf}(z) = \frac{2}{\sqrt{\pi}}\int_0^z e^{-a^2}da$$

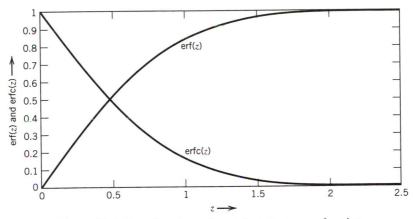

Figure 5A.1 Error function and complementary error function.

The complementary error function of z, $erfc(z) = 1 - erf(z)$.
The error function has the following characteristics:

$$erf(0) = 0$$

$$erf(\infty) = 1$$

$$erf(-z) = -erf(z)$$

Figure 5A.1 plots the values of the error function and the complementary error function of z versus z. Figure 5A.2 is a graph of the complementary error function of z on a logarithmic scale.

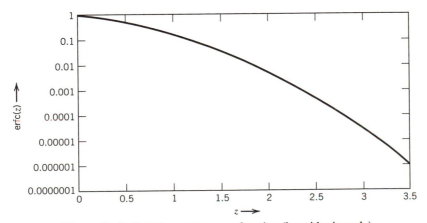

Figure 5A.2 Complementary error function (logarithmic scale).

For small values of z (up to about 0.5), the error function can be approximated as follows:

$$\mathrm{erf}(z) \approx \frac{2}{\sqrt{\pi}} z$$

For large values of z, the complementary error function can be expressed as follows:

$$\mathrm{erfc}(z) \approx \frac{1}{\sqrt{\pi}} \exp(-z^2) \left(\frac{1}{z} - \frac{1}{2z^3} + \frac{1}{2^2 z^5} - \cdots \right)$$

PROBLEMS

5.1 In an experiment to determine the diffusion coefficient of oxygen in nitrogen at 500 K and one atmosphere pressure, the error in temperature measurement is ± 1 K, and the error in the pressure measurement is $\pm 2\%$. Estimate the error limits of the diffusion coefficient based on the expected error limits in temperature and pressure.

5.2 Data on the movement of gases through solids are reported in two different ways: as diffusivities (D) or as permeabilities (P^*). When reported as diffusivities, the concentration term in Fick's first law refers to the concentration of the gas *in the solid.* Thus, for the steady state diffusion of a gas through a layer of solid, Fick's first law is:

$$J_m = -D \frac{\Delta C}{l}$$

where l is the thickness of the solid layer.

When data are reported in terms of *permeability,* Fick's first law has the following form:

$$J_m = -P^* \frac{\Delta p}{l}$$

where P^* is permeability, p is pressure of the gas, and Δp is pressure difference across the solid layer.

(a) What additional information, if any, is needed to convert literature values of permeabilities (P^*) to diffusivities (D)?

(b) The temperature dependence of permeability has the same form as the temperature dependence of diffusivity, that is,

$$P^* = P_0^* \exp\left(-\frac{Q}{RT} \right)$$

Does the Q factor in the permeability equation have the same numerical value as the Q factor in the diffusivity equation? Why?

(c) Carbonated water (seltzer) is now sold in plastic bottles. One of the problems manufacturers of these bottles had to overcome relates to the diffusion of carbon dioxide through the walls of the bottles, with the resulting decrease in carbon dioxide pressure (the seltzer goes "flat"). For the purpose of this problem, assume that a one-liter bottle contains only carbon dioxide gas. The pressure in the bottle is initially (time zero) 3 atms. After storage at constant temperature for 30 days, the pressure is found to have dropped to 2.5 atms. What will be the pressure in the bottle 90 days from time zero? State any assumptions made in your calculation.

5.3 A small quantity of radioactive copper was deposited on the end of a copper bar, which was held at 1073 K for 100 hours. The bar was then sectioned and the activity of each section measured. The resulting data for count rate, in counts per minute (cpm), versus distance from the end of the bar x are as follows

Count Rate (cpm)	$x(10^{-3}$ cm$)$
5000	1
4880	2
4440	3
4040	4
3590	5
3080	6

Plot the data, using appropriate scales on the axes, and determine D, the self-diffusion coefficient of copper at 1073 K.

5.4 A block of copper was joined to a block of α-brass (30 wt % Zn, 70 wt % Cu) with inert markers placed at the interface. After the sample had been held at 785°C for a long time, analysis revealed that the markers were moving at the rate of 2.5×10^{-9} cm/s in the direction of the brass when the experiment was terminated. (Note that the velocity varies with time.) The composition at the markers was 22 mol % zinc. The slope of the concentration curve at the markers $(\partial X_{Zn}/\partial x)$ was 0.89 cm^{-1}. The interdiffusion coefficient at the markers was calculated to be 4.5×10^{-9} cm^2/s.

Calculate the diffusion coefficients of zinc and copper in brass containing 22 mol % zinc at 785°C.

5.5 A solar converter is being created by diffusing phosphorus into a silicon wafer that has been uniformly doped with boron. A p-n junction will be formed at the depth where the phosphorus concentration is equal to the boron concentration.

The boron concentration in the silicon is 10^{16} atoms/cm^3. During the diffusion process, the phosphorus concentration on the surface is held at 10^{20} atoms/cm^3. The diffusion coefficient of phosphorus at the temperature of interest is 10^{-12} cm^2/s.

(a) If the diffusion is carried out for one hour, at what depth will a p-n junction be formed?

(b) What total diffusion time is required to place the p-n junction at double the depth determined in part a?

5.6 This problem is concerned with steady state diffusion in an iron–carbon alloy. The pertinent portion of the Fe–C phase diagram is shown.

A sheet of iron 1 mm thick is treated at 780°C so that one face is maintained at 3.3 atom % carbon and the other face is at zero carbon. Plot the steady state carbon content of the iron as a function of distance through the sheet. Indicate compositions and distances of important points.

For the purpose of this problem, the diffusion coefficients at 780°C are as follows:

For ferrite (α)	$D = 2.5 \times 10^{-6}$ cm²/s
For austenite (γ)	$D = 2.5 \times 10^{-8}$ cm²/s

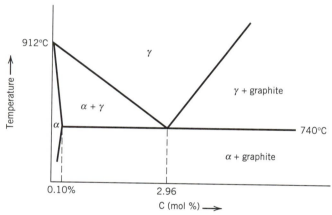

5.7 In a tabulation of diffusion coefficients, the following values are found for Mg^{2+} in MgO:

Temperature (°C)	D (cm²/s)
1450	2.0×10^{-11}
1600	1.3×10^{-10}

What is the diffusion coefficient of Mg^{2+} in MgO at 1500°C? What assumptions are made in your calculation? What sources of error could be introduced by extrapolating the calculation of temperatures beyond the 1450–1600°C range?

5.8 The diffusion coefficient of Mg^{2+} in MgO is determined to be 2.0×10^{-11} cm²/s in a furnace with the temperature controller set at 1450°C. The temperature is measured by a system that specifies an accuracy of $\pm 1\%$ on the indicated temperature. What are the fractional error limits on the diffusion coefficient determined using this apparatus? The activation energy for Mg^{2+} diffusion in MgO is about 76,000 cal/mol.

5.9 As a metal (Me) is heated in oxygen, a layer of MeO grows on the surface. The rate of growth *(dx/dt)* is controlled by the diffusion of some ion through the oxide (either Me^{2+} or O^{2-}).

   (a) Derive an expression for the thickness of the layer in terms of the diffusion coefficient of the diffusing species (D), and the specific volume of the oxide $(\overline{V}_{MeO})$. Assume that the thickness of the layer is zero at the beginning of the process (time = zero).

   (b) Devise an experiment to determine which of the two species is migrating.

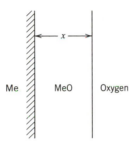

5.10 Helium is to be held in a spherical quartz bulb at 500°C. The helium pressure in the bulb is initially one atmosphere. The outside of the bulb is at zero pressure (a vacuum). Helium is known to diffuse through quartz.

   (a) What is the rate of helium loss, in grams per second, when the pressure inside the bulb is one atmosphere?

   (b) Derive a relationship between the pressure in the bulb as a function of time and the pertinent variables and physical constants (the diffusion coefficient, the temperature, the thickness of the bulb wall, the radius of the bulb, the initial pressure, etc.).

   (c) How long will it be before the pressure inside the bulb drops to half its initial value?

   DATA AND INFORMATION

   The bulb is a spherical shell, 20 cm in diameter.
   The wall thickness of the bulb is 1 mm.
   The diffusion coefficient for helium in quartz at 500°C is 10^{-7} cm²/s
   At 1 atm helium pressure, the solubility of helium in quartz at 500°C is 1.8×10^{-6} g/cm³. The solubility for helium in quartz is linearly proportional to its pressure.
   The atomic weight of He is 4.

5.11 A steel plate is placed in a furnace in which an ammonia (NH_3) atmosphere is maintained at 540°C. Thermal decomposition of NH_3 leads to nitrogen dissolution in the steel plate, with a constant surface concentration of 0.20 wt %. The diffusion constant (D) for nitrogen at this temperature is 7.4×10^{-8} cm²/s.

(a) How long must the dissolution process continue to achieve a nitrogen concentration of 0.044 wt % at a depth of 0.5 mm below the surface?

(b) How long will it take to achieve a concentration of 0.044 wt % at a depth of 1 mm below the surface of the steel plate?

5.12 A piece of n-type silicon contains antimony at a uniform level of 1.44×10^{18} cm^{-3}. It is to be used to generate a p-n junction 5 μm below the surface by diffusing an acceptor dopant from the surface, where its concentration can be considered to be constant at 2×10^{20} cm^{-3}. The diffusion constant for the dopant is 8×10^{-12} cm²/s at the diffusion temperature. What is the diffusion time (in minutes) required to generate the junction?

5.13 When a sample of a stainless steel (an iron–nickel–chromium alloy) is heated in air, an oxide layer forms on its surface. The change in the thickness of the oxide layer can be followed by observing the weight gain of the sample. In a series of experiments it is found that the weight gain (ΔM) can be expressed as a function of time by an equation of the form:

$$(\Delta M)^2 = kt$$

where ΔM = increase in weight
     t = time of exposure to air
     k = a constant

I have found two values of k in the literature:

T (°C)	k (gr/cm²)²/s
600	1.70×10^{-15}
650	3.52×10^{-15}

I would like to know the value of k at 550°C.

(a) How do you expect k to vary with temperature (i.e., what is the functional relationship between k and T)? Why?

(b) Calculate the value of k at 550°C.

5.14 A sheet of polystyrene 1 mm thick is exposed to moist air on one side and dry air on the other. The pressure of water vapor on the moist side is 0.02 atm. The temperature is 25°C.

(a) What is the rate of permeation of water vapor through the polystyrene sheet in cubic centimeter (STP) per centimeter squared per second at steady state?

(b) The polystyrene sheet is coated on one side with a 0.1 mm thick layer of polyvinyl chloride (PVC). What is the water pressure at the interface between the two (P_I) at steady state? (See accompanying diagram.)

(c) What is the rate of permeation of water through the two-layer sheet in cubic centimeters (STP) per centimeter squared per second at steady state?

DATA

Permeation constants at 25°C

Polystyrene $P^* = 1.06 \times 10^{-5}$ cm³(STP)/cm·s·atm
PVC $P^* = 0.27 \times 10^{-5}$ cm³(STP)/cm·s·atm

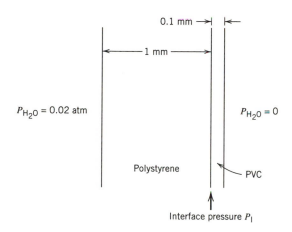

Chapter 6

Transformations

The chapter on surfaces began with the statement, "Everything has to end somewhere." With respect to transformations, we can also observe that "Everything has to start somewhere." The dynamics of phase transformations, especially their initiation, influence many of the changes in the properties of solids that take place during their manufacture.

Phase transformations may be divided into two broad categories: diffusional and displacive (nondiffusional). The former requires movement of atoms by a diffusional process, that is, a process in which atoms move individually, driven by chemical potential gradients. Displacive transformation, on the other hand, involves cooperative movement of atoms in a shearing action during phase transformation. The most striking and commercially important example is the martensite reaction in steel, which accounts for its ability to be hardened. Examples are also to be found in ceramic materials. Good discussions of displacive transformations are to be found in especially the metallurgical literature (Refs. 1 and 2).

Diffusional transformations may be subdivided into two categories: spinodal transformations and those that proceed by nucleation of a new phase, followed by its growth.

6.1 TYPES OF DIFFUSIONAL TRANSFORMATIONS

Gibbs, more than 100 years ago, when considering the transformation of supersaturated solutions in binary systems, observed that there could be two types of transformations: those that are initially small in degree, but large in extent, and those that are initially small in extent, but large in degree. The former are spinodal transformations. To explore the basis for this observation, consider the phase diagram in Figure 6.1, and the corresponding curve for Gibbs free energy versus composition at a temperature of about 800 K in Figure 6.2. The points of inflection in Figure 6.2 where the second derivative of the Gibbs free energy–composition curve is zero, separate the region of positive from negative second derivatives and are called spinodal points. If a solution is cooled rapidly from above the miscibility gap to 800 K, its transformation behavior as it decomposes into α and β phases depends on the overall composition of the material relative to these spinodal points. If the overall composition is inside (between) the spinodal points, the transformation may proceed incrementally over a large region, without the nucleation of a new phase. Outside the spinodal points, nucleation of a new phase is required.

To explain this difference, let us follow, in Figure 6.2, the trajectory of Gibbs free energy during transformations in the two regions: the one with positive second derivative (concave up, point A) and the one with negative second derivative (concave down, point B). In the concave down region, inside the spinodal curve, the decomposition may begin without an initial increase in Gibbs free energy. As the material decomposes into two others, the combined Gibbs free energies of the two resulting phases is always below the Gibbs free energy of the original solution. There is no energy barrier to be overcome. This is the reaction that Gibbs referred to as

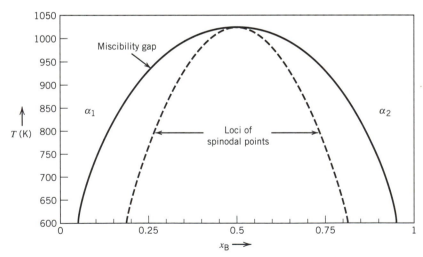

Figure 6.1 Phase diagram showing the miscibility gap and the loci of the spinodal points.

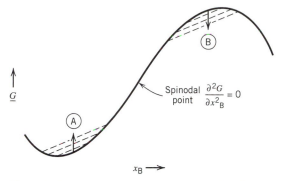

Figure 6.2 Illustration of different paths for changes in $\underline{G}$ on different sides of spinodal point.

small in degree, but large in extent. The reaction can take place over a wide region, but the changes in composition are continuous and small.

In the region of positive second derivative (point A), the overall Gibbs free energy of the material must be increased before transformation to a lower free energy state can occur. The sum of the Gibbs free energies of the transformed material is initially higher than the Gibbs free energy of the solution. As the reaction proceeds, the sum of the two will eventually be less than that of the solution, but there is an energy barrier to be overcome in the overall process. These reactions proceed by the process of nucleation and growth, and are, in Gibbs's terms, small in extent, but large in degree.

The difference in transformation kinetics between the two can be visualized nicely using a mechanical analogue (Ref. 3). Consider the rectangular block in Figure 6.3. The stable equilibrium of the block is on its side, portrayed at the right. The position on the left, on its end, is labeled "metastable" because a certain amount of work is

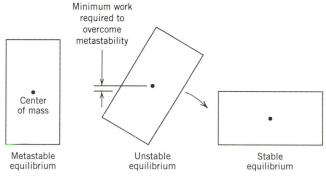

Figure 6.3 Mechanical analogue of the nucleation process. (From Ref. 3.)

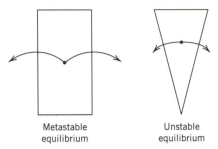

Metastable	Unstable
equilibrium	equilibrium

Figure 6.4 Illustration of the difference between metastable equilibrium and unstable equilibrium. (From Ref. 3.)

first required to lift its center of gravity as the block falls to its stable equilibrium position.

Consider, now, the difference between the rectangular block and the triangular block in Figure 6.4. The triangular block, balanced on its tip, represents unstable equilibrium because any slight perturbation from the vertical position will cause the block to fall to its stable position (on its side) without the addition of any work to start the process. The energy path followed by the rectangular block is analogous to transformations characterized by nucleation. The energy path followed by the triangular block is analogous to spinodal transformations. Nucleation is discussed first in this chapter. Section 6.8 deals with spinodal transformation.

6.2   HOMOGENEOUS NUCLEATION

To begin the discussion of phase changes of the second kind (nonspinodal), consider the Gibbs free energy changes during the solidification of a pure material. At temperatures below a material's melting point (T_m), the driving force for solidification is the difference in Gibbs free energy (ΔG) between the liquid and the solid. If we assume that the heat capacities of the liquid and solid are equal, then the molar enthalpy and molar entropy of solidification will each remain constant as a function of temperature, and $\Delta \underline{G}$ can be calculated as follows:

$$\text{liquid} \rightarrow \text{solid}$$

$$\Delta \underline{G} = \Delta \underline{H} - T \, \Delta \underline{S}$$

Note that $\Delta \underline{H} = -L$, where L is the latent heat of fusion.

$$\Delta \underline{G} = -L + T \frac{L}{T_m}$$

$$\Delta \underline{G} = \frac{L}{T_m} (T - T_m) \tag{6.1}$$

When a spherical particle of solid of radius r is formed, the change in Gibbs free energy is the volume of the particle multiplied by the volumetric Gibbs free energy change, $\Delta \underline{G}_v$.

$$\Delta G_{vol} = \tfrac{4}{3}\pi r^3 \Delta \underline{G}_v$$

where $\Delta \underline{G}_v$ is the Gibbs free energy change per unit volume,

$$\Delta \underline{G}_v = \frac{1}{\underline{V}}\frac{L}{T_m}(T - T_m) \tag{6.2}$$

$$\Delta G_{vol} = \frac{4}{3}\pi r^3 \frac{1}{\underline{V}}\frac{L}{T_m}(T - T_m) \tag{6.3}$$

But when the particle of radius r is formed, there is another energy term to be considered, the surface energy. The surface energy of the particle is

$$\Delta G_s = 4\pi r^2 \gamma \tag{6.4}$$

where $\gamma = \gamma_{s-1}$, the surface energy between solid and liquid.

The sum of the two energy terms is:

$$\Delta G_t = 4\pi r^2 \gamma + \tfrac{4}{3}\pi r^3 \Delta \underline{G}_v \tag{6.5}$$

The first of these terms involves the increase in energy required to form a new surface. The second term is negative and represents the decrease in Gibbs free energy upon solidification. Because the first is a function of the second power of the radius, and the second a function of the third power of the radius, the sum of the two increases, goes through a maximum, and then decreases (Figure 6.5). The radius at which the Gibbs free energy curve is at a maximum is called the critical radius r^*, for a nucleus of solid in liquid. The driving force of the Gibbs free energy will tend to cause a particle with a smaller radius than r^* to decrease in size. This is a particle of subcritical size for nucleation. A viable nucleus is one with radius greater than or equal to r^*. The critical Gibbs free energy corresponding to the radius r^* is ΔG^*. In terms of physical parameters, these terms can be shown to be:

$$r^* \quad \text{when} \quad \left(\frac{\partial \Delta G_t}{\partial r}\right)_T = 0 = 8\pi r \gamma + 4\pi r^2 \Delta \underline{G}_v \tag{6.6}$$

$$r^* = -\frac{2\gamma}{\Delta \underline{G}_v} \tag{6.7}$$

$$\Delta G^* = \frac{16}{3}\frac{\pi \gamma^3}{\Delta \underline{G}_v^2} \tag{6.8}$$

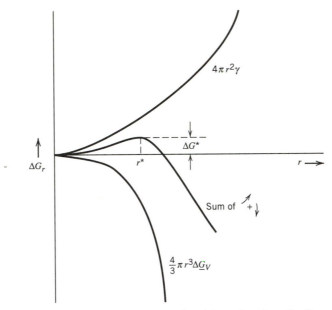

Figure 6.5 Gibbs free energy of nuclei as a function of radius.

6.3   SIZE DISTRIBUTION OF NUCLEI

One may well ask why nucleation takes place at all. How does a nucleus of size greater than $r*$ form at any time? Based on the principles discussed in Chapter 2 (Statistical Thermodynamics), we can calculate the probability that a particle of radius r will exist at a given temperature, and that probability will be greater than zero. Let the particle with radius r have a Gibbs free energy of formation ΔG_r, calculated according to Eq. 6.5. Consider now the entropy of mixing of these particles (numbering n_r) with atoms of the liquid (i.e., particles having atomic radius: numbering n_0). To minimize the Gibbs free energy of the combination of the two:

$$\Delta S_{\text{mix}} = k \ln \frac{(n_0 + n_r)!}{n_0! n_r!} \qquad (6.9)$$

$$\Delta G = n_r \Delta G_r - kT[(n_0 + n_r)\ln(n_0 + n_r) - n_0 \ln n_0 - n_r \ln n_r]$$

At the equilibrium value of n_r, $(\partial \Delta G_r / \partial n_r)_T = 0$ and

$$\Delta G_r + kT \left(\ln \left[\frac{n_r}{n_0 + n_r} \right] \right) = 0$$

Noting that $n_0 \gg n_r$,

$$\frac{n_r}{n_0} = \exp\left(-\frac{\Delta G_r}{kT}\right) \tag{6.10}$$

One could, based on Eq. 6.10, calculate the concentration of solid nuclei of radius r in a liquid at its melting point. In this case, the value of ΔG_v is zero because the two phases, liquid and solid, are in equilibrium. Only the surface energy term of the equation comes into play. As an example, let us calculate the distribution of nuclei in nickel at its melting point, 1725 K.

For nickel: $T_m = 1725$ K

$$\frac{L}{T_m} = 10 \text{ J mol}^{-1} \text{ K}^{-1}$$

$$\gamma = 0.25 \text{ J/m}^{-2}$$

$$\underline{V} = 7 \text{ cm}^3/\text{mol}$$

$$n_0 = \frac{N_A}{\underline{V}} = 8.6 \times 10^{22} \text{ cm}^{-3}$$

From Eq. 6.5:

$$\Delta G_r = 4\pi r^2 \gamma + \tfrac{4}{3}\pi r^3 \Delta \underline{G}_v$$

At the melting point, T_m, $\Delta \underline{G}_v = 0$ (see Eq. 6.2)
Thus:

$$\Delta G_r = 4\pi r^2 \gamma$$

As an example, let us calculate the concentration of clusters of radius 7 Å at the melting point:

$$\Delta G_r = 4\pi(7 \times 10^{-10})^2(0.25) = 1.54 \times 10^{-18} \text{ J}$$

$$n_r = n_0 \exp\left(-\frac{\Delta G_r}{kT}\right)$$

$$n_r = 8.6 \times 10^{22} \exp\left[-\frac{1.5 \times 10^{-18}}{(1.38 \times 10^{-23})(1725)}\right]$$

$$n_r = 7.3 \times 10^{-6} \text{ clusters/cm}^3$$

By repeating the calculation for a series of cluster radii, we obtain the results in Table 6.1.

Table 6.1 Calculated Concentration of Solid Clusters in Liquid Nickel at the Melting Temperature as a Function of Cluster Size

$r(\mathring{A})$	$n_r(\text{clusters/cm}^3)$
5	4×10^9
6	2×10^2
7	7.3×10^{-6}
8	1.8×10^{-14}
10	4.4×10^{-35}

The results of the calculations in Table 6.1 show that the concentration of solid nuclei depends very strongly on cluster radius. This strong dependence enables us to speak of a "maximum" cluster radius at a given temperature (see Figure 6.6). Based on Eqs. 6.10 and 6.5, there is no true maximum cluster radius at a specified temperature. For every cluster size, there is a calculated concentration level. But a consideration of Table 6.1 should convince us that when concentrations of nuclei fall below one per cubic centimeter, we enter a size regime in which concentrations fall very rapidly as a function of cluster radius. We may, for our purposes, pick a reasonable concentration of nuclei, say one per cubic centimeter, and specify the cluster radius that exists at that concentration to be the maximum size. At the melting point of nickel, that calculated maximum radius would be about 6.3 Å.

6.4   SUPERCOOLING

Based on Eqs. 6.7, 6.8, and 6.10, we should be able to calculate the temperature at which a liquid will start to solidify through the process of homogeneous nucleation. As an example, let us calculate the probability of finding nuclei of critical size in pure nickel that has been supercooled 10 K below its melting point. Using the phys-

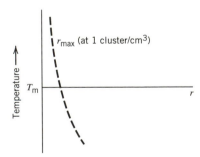

Figure 6.6 Maximum cluster radius r_{max} as a function of temperature.

ical properties of nickel from Section 6.3, we express the critical radius for such a nucleus as follows:

$$r^* = -\frac{2\gamma}{\dfrac{L}{T_m}\dfrac{1}{\underline{V}}(T - T_m)} = -\frac{(2)(0.25)}{10\left(\dfrac{1}{7 \times 10^{-6}}\right)(-10)}$$

$$r^* = 350\ \text{Å}$$

The Gibbs free energy of this nucleus is:

$$\Delta G^* = \frac{16}{3}\frac{\pi\gamma^3}{\left[\dfrac{L}{T_m}\dfrac{1}{\underline{V}}(T_m - T)\right]^2}$$

$$\Delta G^* = 1.3 \times 10^{-15}\ \text{J}$$

Based on Eq. 6.10, the concentration of such nuclei is

$$n_r = n_0 \exp\left[-\frac{\Delta G^*}{kT}\right] = 8.6 \times 10^{22}\ \exp\left[-\frac{1.3 \times 10^{-15}}{(1.38 \times 10^{-23})(1715)}\right]$$

$$n_r = (8.6 \times 10^{22})(10^{-24,000})$$

$$n_r \approx 10^{-24,000}\ \text{cm}^{-3}$$

We may thus safely conclude that homogeneous nucleation does not take place in nickel at 10 K below its melting point, because the concentration of nuclei of the critical size is so low.

Using the same equations, we may calculate the maximum cluster size in liquid nickel as a function of temperature (Figure 6.6), noting that by ''maximum'' we mean the cluster size that is present in a concentration of at least one per cubic centimeter. The critical radius of nuclei as a function of temperature may be

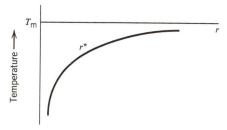

Figure 6.7 Critical radius r^* as a function of temperature.

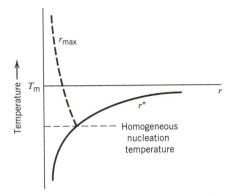

Figure 6.8 Superposition of Figures 6.6 and 6.7.

calculated from Eq. 6.7. The results are shown schematically in Figure 6.7. Figure 6.8 is a superposition of Figures 6.6 and 6.7. The intersection of the critical radius curve of Figure 6.7 with the maximum radius curve of Figure 6.6 should be at the temperature of homogeneous nucleation. We can calculate this temperature for pure nickel by repeating the calculation above for various degrees of supercooling. The results of such a calculation are presented in Table 6.2.

Based on Table 6.2, homogeneous nucleation should take place at about 340–350 K of supercooling. In fact, the maximum degree of supercooling for nickel is 319 K, observed in an experiment on the solidification of very fine droplets of liquid nickel.

6.5 HETEROGENEOUS NUCLEATION

Although it is possible to achieve supercooling levels exceeding 300 K in nickel, substantial supercooling of this magnitude takes place only under very carefully controlled experimental conditions. Under practical solidification conditions, super-

Table 6.2 Critical Radius and Concentration of Nucleii of Critical Radius as a Function of Temperature for Nickel[a]

$T - T_m$	r^* (Å)	ΔG (J)	n_r^* (cm^{-3})
10	350	1.3×10^{-15}	$10^{-24,000}$
100	35	1.3×10^{-17}	10^{-253}
300	11.7	1.45×10^{-18}	7×10^{-10}
325	10.7	1.24×10^{-18}	1.4×10^{-5}
340	10.3	1.12×10^{-18}	2.8×10^{-3}
400	9.0	8.2×10^{-19}	3700

[a]Actually: $T - T_m$ is observed for Ni at approximately 319 K.

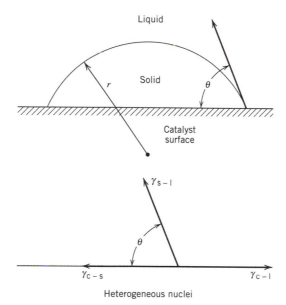

Figure 6.9 Catalysis of a nucleus on a surface.

cooling of only a few degrees is observed because nuclei of the solid can be formed on surfaces that catalyze solidification, such as inclusions in the material being solidified, the walls of the container in which it is being held, or the surfaces of the casting molds. To catalyze solidification, the nucleus of solid must wet the catalyst to some extent. A nucleus catalyzed on a surface is shown schematically in Figure 6.9.

The analysis of the energies involved in heterogeneous nucleation follows the same method as the one used for homogeneous nucleation in Section 6.2. In the case of a heterogeneous nucleus in the form of a spherical cap (Figure 6.9), the surface energy term involves the surface energy between the solid nucleus and the liquid, and the change in surface energy of the catalyst surface as it is coated by the nucleus. The surface energy term is derived as follows:

$$\text{solid–liquid surface} = 2\pi r^2(1 - \cos\theta) \tag{6.11}$$

$$\text{catalyst–solid surface} = \pi r^2(1 - \cos^2\theta) \tag{6.12}$$

where r is the radius of curvature of the nucleus. Then we write

$$\Delta G_{\text{surface}} = 2\pi r^2(1 - \cos\theta)\gamma_{s-1} + \pi r^2(1 - \cos^2\theta)(\gamma_{c-s} - \gamma_{c-1}) \tag{6.13}$$

where γ_{s-1} = solid–liquid interfacial energy
γ_{c-s} = solid–catalyst interfacial energy
γ_{c-1} = liquid–catalyst interfacial energy

The terms involving the interactions between the catalyst surface and the liquid and the solid can be expressed in terms of the solid–liquid interfacial energy by noting the relationships among them (Section 4.10):

$$\gamma_{c-l} = \gamma_{c-s} + \gamma_{s-l}(\cos \theta)$$

The volumetric Gibbs free energy change is the product of the volume of the cap and $\Delta \underline{G}_v$, the specific Gibbs free energy change. That volume, in terms of its radius of curvature and contact angle, is:

$$V = \frac{4}{3}\pi r^3 \left\{ \frac{(2 + \cos \theta)(1 - \cos \theta)^2}{4} \right\} \tag{6.14}$$

or

$$\Delta G_{\text{volumetric}} = \tfrac{4}{3}\pi r^3 \Delta \underline{G}_v f(\theta) \tag{6.15}$$

where

$$f(\theta) = \left\{ \frac{(2 + \cos \theta)(1 - \cos \theta)^2}{4} \right\} \tag{6.16}$$

Following the method of Section 6.2 yields the following conclusion:

$$r^* = - \frac{2\gamma_{s-l}}{\Delta \underline{G}_v} \quad \text{and} \quad \Delta G^* = \frac{16}{3}\frac{\pi \gamma_{s-l}^3}{\Delta \underline{G}_v^2} f(\theta) \tag{6.17}$$

It is particularly important to note that the critical radius of curvature, r^*, does not change when the nucleation becomes heterogeneous. The critical Gibbs free energy, ΔG^*, however, is strongly influenced by the wetting that occurs at the surface of the material that catalyzes the nucleation. A lower value of ΔG^* means a lower activation energy to be overcome in nucleation; that is, nucleation takes place more easily. The magnitude of the effect can be appreciated by considering values of $f(\theta)$, shown in Table 6.3. It can be shown that $f(\theta)$ is the ratio of the volume of the

Table 6.3 Values of $f(\theta)$ in Eq. 6.17, Indicating the Extent to Which the Activation Energy Is Reduced by Wetting of the Nucleus

θ	$f(\theta)$
90°	0.50
60°	0.16
30°	1.3×10^{-2}
10°	7.0×10^{-4}

heterogeneous nucleus (the cap) to the volume of the sphere with the same radius of curvature.

Figure 6.10, a graph of ΔG as a function of radius of curvature of the nucleus, shows the effect of wetting on the critical Gibbs free energy to be overcome for the nucleus to form.

The critical Gibbs free energy for nucleation depends on the nucleus volume. This can be demonstrated by considering a nucleus having the shape of spherical cap with radius of curvature r. The Gibbs free energy of the nucleus depends on the interfacial energy and the volumetric Gibbs free energy change as follows:

$$\Delta G_r = \alpha r^2 \gamma + \beta r^3 \Delta \underline{G}_v \qquad (6.18)$$

The parameters α and β are determined by the particular geometry of the nucleus. The surface energy term, γ, is an average surface energy for the nucleus determined according to the geometrical factors.

The volume of the nucleus is βr^3. To determine r^*,

$$\left(\frac{\partial \Delta G_r}{\partial r} \right)_T = 0$$

$$2\alpha \gamma r^* + 3\beta r^{*2} \Delta \underline{G}_v = 0$$

$$r^* = -\frac{2\alpha}{3\beta \Delta \underline{G}_v} \gamma \quad \text{or} \quad \alpha = -\frac{3\beta \Delta \underline{G}_v}{2\gamma} r^* \qquad (6.19)$$

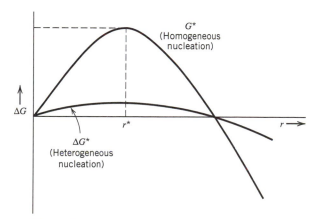

Figure 6.10 Plot of ΔG versus r for homogeneous nucleation and an example of heterogeneous nucleation.

Substituting in Eq. 6.18, we have

$$\Delta G_r^* = -\frac{3\beta\underline{\Delta G}_v r^*}{2}(r^*)^2 + \underline{\Delta G}_v\beta(r^*)^3$$

$$\Delta G_r^* = -\tfrac{1}{2}\beta(r^*)^3\underline{\Delta G}_v$$

$$\Delta G_r^* = -\tfrac{1}{2}V^*\underline{\Delta G}_v \tag{6.20}$$

where V^* is the volume of the critical nucleus.

From this analysis, it is apparent that any factors that reduce the volume of the nucleus reduce the critical Gibbs free energy of formation of that nucleus, making nucleation more probable.

6.6   RATE OF NUCLEATION

The rate of nucleation in a transformation is determined by the concentration of nuclei of the critical size and the rate at which they are "activated" through the addition of atoms or molecules to their surfaces. This may be expressed as follows:

$$\dot{N} = v n_s n^* \tag{6.21}$$

where n^* = concentration of critical nuclei
n_s = number of atoms or molecules on the nucleus surface
v = collision frequency of molecules with nuclei

In the case of the formation of nuclei from a vapor (ideal gas), the collision frequency is given by the Langmuir equation (Section 2.12). The rate of nucleation is then:

$$\dot{N} = \frac{\alpha_c P}{(2\pi mkT)^{1/2}} A^* n_0 \exp\left(-\frac{\Delta G^*}{kT}\right) \tag{6.22}$$

where A^* is the area of the critical nucleus.

In the case of transformation in condensed phases, the collision frequency at the nucleus interface may be expressed as follows:

$$v = v_0 \exp\left(-\frac{\Delta G_M}{kT}\right) \tag{6.23}$$

where v_0 is the jump frequency of the molecules or atoms at the surface of the nucleus and ΔG_M is the activation energy for the movement to the nucleus.

The nucleation rate would have the form:

$$N = v_0 n_s n_0 \exp\left(-\frac{\Delta G^*}{kT}\right) \exp\left(-\frac{\Delta G_M}{kT}\right) \tag{6.24}$$

The first exponential term in Eq. 6.24 increases as the temperature decreases below the equilibrium temperature for the reaction. To use solidification as an example,

$$\Delta G^* = \frac{16}{3} \frac{\pi\gamma^3}{\left[\dfrac{L}{T_m}\dfrac{1}{V}(T - T_m)\right]^2} \tag{6.25}$$

At the melting temperature, T_m, ΔG^* is infinite and $\exp(-\Delta G^*/kT)$ is zero. As the temperature drops, the value of this exponential term increases.[1] The value of the second exponential term, $\exp(-\Delta G_M/kT)$, decreases as the temperature decreases, assuming that the activation energy term, ΔG_M, remains constant. The product of the two goes through a maximum as the temperature drops farther below the equilibrium temperature. This is illustrated in Figure 6.11. Because the nucleation rate passes through a maximum, we have the possibility of cooling a material rapidly enough to suppress the transformation altogether. This would be accomplished by passing through the temperature of the maximum nucleation rate before the equilibrium phase nucleates.

---

[1]It is interesting to note that the term $\exp(-\Delta G^*/kT)$ does not increase monatonically as temperature decreases below the equilibrium temperature. It can be shown to have a maximum as follows:

$$\frac{\Delta G^*}{kT} = \frac{16}{3} \frac{\pi\gamma^3}{(L^2/T^2)(1/V^2)k} \left(\frac{1}{T}\right) \left(\frac{1}{T - T_m}\right)^2 = K \left(\frac{1}{T}\right) \left(\frac{1}{T - T_m}\right)^2$$

Maximum at

$$\frac{d[\exp(-\Delta G_m/kT)]}{dT} = 0$$

$$\exp\left[-K \left(\frac{1}{T}\right) \left(\frac{1}{T - T_m}\right)^2\right] \left[\frac{2}{T}\left(\frac{1}{T - T_m}\right)^3 - \left(\frac{1}{T - T_m}\right)\left(\frac{1}{T^2}\right)\right] = 0$$

$$T = \frac{T_m}{3} \text{ at maximum}$$

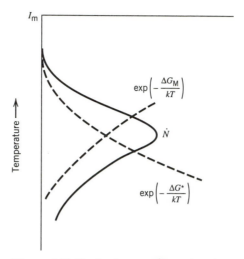

Figure 6.11 Nucleation rate $\dot{N}$ as a function of temperature.

6.7   GROWTH OF TRANSFORMED PHASE

The rate of growth of a transformation product is determined by the driving force for the transformation and the frequency with which molecules successfully make the transition from the reactant phase to the product phase. To use solidification as an example, the *driving force* is the negative of the ΔG of solidification:

$$-\Delta \underline{G} = \frac{L}{T_m} (T_m - T) \qquad (6.26)$$

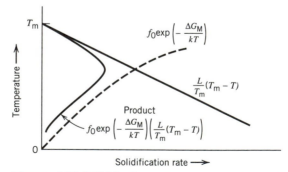

Figure 6.12 Solidification rate as a function of temperature.

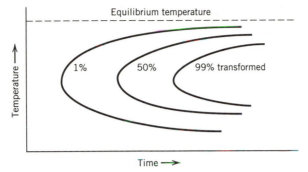

Figure 6.13 A TTT diagram.

The jump frequency across the liquid–solid interface has a temperature dependence of the form:

$$f = f_0 \exp\left(-\frac{\Delta G_M}{kT}\right) \tag{6.27}$$

where ΔG is the activation energy for movement across the liquid–solid interface.
The product of the two is:

$$\text{solidification rate} = f_0 \exp\left(-\frac{\Delta G_M}{kT}\right)\left(\frac{L}{T_M}\right)(T_m - T) \tag{6.28}$$

As the temperature decreases, the driving force increases but the jump frequency decreases. These two opposing dependencies can produce a maximum in the rate of growth as a function of temperature, as illustrated in Figure 6.12.

The temperature dependencies of both nucleation of a new phase and its rate of growth result in a strong temperature dependence of transformation rate. Results of transformation studies are often presented in the form of diagrams in which the time required to transform a specific amount of the material is plotted against the temperature of transformation. A schematic example is shown in Figure 6.13. These curves are often referred to as time–temperature–transformation (TTT) curves, or "C" curves because their shape is in the form of the letter C.

The TTT curve of a transformation indicates that it may be possible to cool a material rapidly enough to slide past the nose of the curve, that is, to avoid the transformation and to arrive at a temperature where the transformation rate is very slow. This accounts for the existence of nonequilibrium structures, such as silicate glasses, and amorphous metals.

6.8   SPINODAL TRANSFORMATIONS

Two different types of diffusive transformations were discussed in Section 6.1—spinodal and those that are initiated by nucleation. The difference between the two was shown in graphical form in Figures 6.2–6.4. Nucleation was discussed in Section 6.2. Now we discuss the principles of decomposition of homogeneous solutions by the spinodal mechanism.

Recognize at the outset that we are discussing the decomposition of a supersaturated solution into its equilibrium phases. The solution became supersaturated, presumably, by being cooled into a temperature–composition region inside a miscibility gap. In the binary system we are considering, the solution will eventually decompose into a mixture of two phases.

Consider, first, the change of Gibbs free energy of the homogeneous, supersaturated solution as it undergoes composition fluctuations. For the purpose of our analysis, let the solution of overall composition c_0 undergo local fluctuations of $\pm\delta c$; that is, it will split up into two regions of composition $c_0 + \delta c$ and $c_0 - \delta c$. We can expand the Gibbs free energy of the solution as a function of composition around the composition c_0 by a Taylor series:

$$G_{c_0 + \delta} = G_{c_0} + (\pm\delta c)G'_{c_0} + \tfrac{1}{2}(\pm\delta c)^2 G''_{c_0}$$

where

$$G'_{c_0} = \left(\frac{\partial G}{\partial c}\right)_{c_0} \quad \text{and} \quad G''_{c_0} = \left(\frac{\partial G^2}{\partial^2 c}\right)_{c_0}$$

The change in Gibbs free energy accompanying the composition fluctuation is:

$$\Delta G = G_{c_0 \pm \delta c} - G_{c_0}$$

$$\Delta G = \tfrac{1}{2}[(\delta c)G'_{c_0} + \tfrac{1}{2}(\delta c)^2 G''_{c_0}] + \tfrac{1}{2}[(-\delta c]G'_{c_0} + \tfrac{1}{2}(-\delta c)^2 G''_{c_0}]$$

$$\Delta G = \tfrac{1}{2}(\delta c)^2 G''_{c_0} \tag{6.29}$$

If the second derivative of G with respect to composition is positive, then ΔG accompanying the fluctuation is positive, and the fluctuation will tend to collapse. This tells us that the solution is in a *metastable,* but not *unstable,* state. The decomposition of such a solution requires that nuclei of a new phase be formed as discussed in Section 6.2.

If, however, the second derivative of G with respect to composition is negative, then the ΔG accompanying the fluctuation is negative, and the fluctuations will tend to intensify. The solution is unstable. This is the spinodal decomposition case. The differences in composition profiles during the development of a new phase from the supersaturated solution are shown schematically in Figure 6.14 (Ref. 3) for the two cases.

As we have so far developed the basis for spinodal decomposition, we could

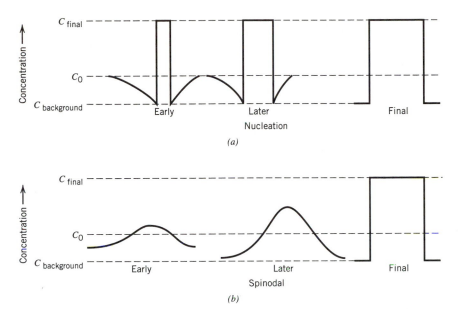

Figure 6.14 Schematic evolution of concentration profiles to illustrate the difference between the spinodal mechanism (b) and nucleation and growth (a).

conclude that there is no size limit or dimensional scale that characterizes the new phase. This is not the case, however. A wavelength or characteristic dimension is observed in spinodal decompositions. The explanation for this phenomenon lies in the observation that the Gibbs free energy of an atom in a *concentration gradient* is not the same as the Gibbs free energy of that atom in a solution of *uniform* composition (Ref. 4). Hillert showed this to be the case using a regular solution model. A rigorous derivation of the relationship (Ref. 5) yields, for a one-dimensional case:

$$G = N_V \int_V \left[\underline{G}_c + \kappa \left(\frac{dc}{dx} \right)^2 \right] dV \qquad (6.30)$$

where N_V = number of atoms per unit volume
     $\underline{G}(c)$ = Gibbs free energy (per atom) of a homogeneous
        solution of composition c
    dc/dx = concentration gradient
      κ = a constant that is positive when two phases tend to
        separate

Equation 6.30 tells us that the Gibbs free energy of a volume V of solution in a concentration gradient has two parts: the Gibbs free energy the volume would have if it existed as a part of a homogeneous solution of uniform composition, and another part that depends on the concentration gradient at that point. This last term is called the *gradient energy*. The concept of "gradient energy" is not intuitively obvious. Its plausibility can be demonstrated by comparing the energy associated with the bond energy of an atom with its nearest neighbors in a uniform solution, and in a concentration gradient. An example of such a calculation is presented in Appendix 6A.

If we expand $\underline{G}(c)$ in Eq. 6.30 in a Taylor series, and note that the odd terms will vanish in an isotropic medium, then the equation becomes:

$$\Delta G = N_V \int_V \left[\frac{1}{2} G''_{c_o} (\delta c)^2 + \kappa \left(\frac{dc}{dx} \right)^2 \right] dV \tag{6.31}$$

where G''_{c_o} is the second derivative of G with respect to concentration at C_o.

In Eq. 6.31, the value of G''_{c_o} is inherently negative inside the spinodal curve. The decomposition as a spinodal can proceed only if the overall ΔG is negative, that is, if[2]:

$$\frac{1}{2} \left| G''_{c_o} \right| (\delta c)^2 > \kappa \left(\frac{dc}{dx} \right)^2 \tag{6.32}$$

Based on Eq. 6.32, there is a minimum magnitude of fluctuation, δc, below which it is unstable. One solution of Eq. 6.31 has the form of a wave. Consider a composition fluctuation of size δc in the x direction expressed as a wave:

$$c - c_o = \delta c = A \cos \beta x \tag{6.33}$$

where $\beta = 2\pi/\lambda$.

By substituting in Eq. 6.33 in Eq. 6.31, and noting that $\int_0^\pi \sin^2 ax\ dx = \int_0^\pi \cos^2 ax\ dx = \pi/2$,

$$\frac{\Delta G}{V} = \frac{\pi A^2}{A} [G''_{c_o} + 2\kappa\beta^2] \tag{6.34}$$

Within the spinodal region the term G''_{c_o} is negative. The homogeneous solution of interest becomes unstable; that is, it will decompose as a spinodal, when

---

[2]There can be other terms in the equation, such as the one representing elastic energy, if region of composition fluctuation is crystallographically coherent with the lattice of the solution but has different dimensions. This is discussed in Ref. 5. The conclusions reached relative to the existence of a minimum length of the new phase do not depend on this coherency energy, although it is important.

$\Delta G < 0$. The critical value of β, β_c, at which $\Delta G = 0$ is:

$$\beta_c = \left[-\frac{1}{2\kappa} G_{c_o}'' \right]^{1/2} \tag{6.35}$$

At values of β greater than β_c ($\lambda < \lambda_c$), the spinodal decomposition will not proceed. The critical wavelength, λ_c is

$$\lambda_c = \left[\frac{-8\pi^2\kappa}{G_{c_o}''} \right]^{1/2} \tag{6.36}$$

At values of λ less than λ_c, spinodal fluctuations will decay. The solution can still decompose to its equilibrium state, but not by the spinodal mechanism.

We can extend the treatment above to a consideration of the kinetics of the spinodal transformation (Refs. 3 and 5). The amplitude of a sinusoidal fluctuation can be expressed as follows:

$$A(\beta, t) = A(\beta, 0) \, \exp[R(\beta)t] \tag{6.37}$$

The amplification factor, $R(\beta)t$, is:

$$R(\beta)t = -\frac{B\beta^2}{N_v} [G_{c_o}'' + 2\kappa\beta^2] \tag{6.38}$$

where B is the mobility, $A(\beta, t)$ is the amplitude of the fluctuation with wavelength β at time t, and $A(\beta, 0)$ is the initial amplitude at $t = 0$.

Based on Eq. 6.38, when $|G_{c_o}''| > 2\kappa\beta^2$, $R(\beta)$ is positive. Remember that G_{c_o}'' is

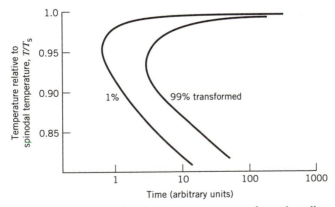

Figure 6.15 Schematic time–temperature–transformation diagram for spinodal decomposition. (From Ref. 7.)

negative in the spinodal region. If the opposite is true ($|G_{c_0}''| < 2\kappa\beta^2$), $R(\beta)$ will be negative, and the fluctuation will decay with time.

The amplification factor, $R(\beta)$, varies with wave number β and shows a maximum value, β_m when $\lambda = \sqrt{2}\lambda_c$, where λ_c is the critical wavelength (Eq. 6.36). The extent of a spinodal transformation can be expressed as a TTT diagram (or C curve) as in the case of nucleation and growth (Figure 6.13). A representative sample of such a curve is shown in Figure 6.15(7).

Spinodal transformations are observed in metallic, ceramic (Refs. 3 and 4), and polymer systems (Ref. 8).

REFERENCES

1. Porter, D. A., and Easterling, K. E., *Phase Transformations in Metals and Alloys,* second edition, Chapman and Hall, *London,* 1992.
2. Massalski, T. B., *Phase Transformations,* pp 433–483, American Society for Metals, 1968.
3. Cahn, J. W., *Trans. A.I.M.E., 242,* 166 (1968).
4. Hillert, M., *Acta Met., 9,* 525 (1961).
5. Cahn, J. W., and Hilliard, J. E., *J. Chem. Phys., 28,* 258 (1958).
6. Kingery, W. D., Bowen, H. K., and Uhlmann, D. R., *Introduction to Ceramics,* second edition. Wiley, New York, 1976.
7. Huston, E. L., Cahn, J. W., and Hilliard, J. E., *Acta Met., 14,* 1053 (1966).
8. "Structure and Properties of Polymers," *Materials Science and Technology,* Vol. 12, VCH Publishers,*Weinheim,* Germany, 1993.

APPENDIX 6A

Gradient Energy in a Regular Solution

Equation 6.30 indicates the Gibbs free energy of an atom in a *concentration gradient* is not the same as the Gibbs free energy of that atom in a solution of *uniform* composition. According to the equation, a gradient energy term, proportional to dc/dx must be added to the Gibbs free energy term for the atom in a uniform solution.

$$G = N_V \int_V \left[\underline{G}_c + \kappa \left(\frac{dc}{dx} \right)^2 \right] dV \qquad (6.30)$$

where $\underline{G}(c)$ is the Gibbs free energy (per atom) of a homogeneous solution of composition c, and dc/dx is the concentration gradient.

The purpose of this calculation is to demonstrate that a gradient energy term will arise when a regular solution model of atomic interactions (nearest neighbor) is used to describe the energy of an atom in a concentration gradient. Only the energy term of the regular solution model is needed.

In a regular solution, the energy of an atom in a uniform binary solution of composition c with coordination number Z is $\omega c(1 - c)$, where ω is $Z(E_{AB} - \frac{1}{2}(E_{AA} + E_{BB}))$. For convenience, let us write this as $Zc(1 - c)\Delta E$, where ΔE is $[E_{AB} - \frac{1}{2}(E_{AA} + E_{BB})]$.

Let us consider an atom in a close-packed plane in a close-packed structure. Its overall coordination number is 12. In the close-packed plane there are six nearest neighbors, and there are three nearest neighbors in each of the neighboring planes. The energy of this atom will be the sum of its interaction with the 12 nearest neighbors, six in its plane and three each in the neighboring planes.

If the atom exists on a plane where the concentration is c_o, the next plane (to its right) will have a concentration of $c_o + (dc/dx)\Delta x$, where Δx is the interplanar spacing and dc/dx is the concentration gradient. The plane to the left will have a concentration of $c_o - (dc/dx)\,\Delta x$. For the purpose of this calculation, let $\Delta c = (dc/dx)\Delta x$. The energy of the atom in the concentration gradient is:

$$E = 6[c(1 - c)\Delta E] + 3[(c + \Delta c)(1 - (c + \Delta c))]\Delta E$$
$$+ 3[(c - \Delta c)(1 - (c - \Delta c))]\Delta E$$
$$E = 6[c(1 - c)\Delta E] + 6[c(1 - c) + (\Delta c)^2]\Delta E$$
$$E = 12[c(1 - c)\Delta E] + 6(\Delta c)^2\Delta E$$
$$E = 12[c(1 - c)\Delta E] + 6\left(\frac{dc}{dx}\right)^2 (\Delta x)^2\Delta E$$

This first term in the energy equation, $12c(1 - c)\Delta E$, is the energy the atom would have if it were in a uniform solution of composition c. The second term, $6(dc/dx)^2(\Delta x)^2\Delta E$, is the gradient energy term.

PROBLEMS

6.1 For pure, liquid copper at its melting point, what is the Gibbs free energy of formation of a spherical crystalline cluster of solid (a nucleus of solid) of radius 5, 7, 10Å?

Estimate the concentration of such nuclei (r = 5, 7, 10 Å).

For copper:

Melting temperature = 1083°C

γ_{s-l} = 200 ergs/cm^2

Heat of fusion = 3120 cal/mol

Assume that the molar volume is 7.0 cm^3/mol for both solid and liquid.

6.2 In experiments on homogeneous nucleation, it has been found that solidification of many pure metals can be suppressed to a temperature where $\Delta G^*/kT$, the exponential term in the nucleation equation, reaches a value of about 30. Assuming that copper is such a metal, estimate the value of the liquid–solid interfacial energy using the following data.

DATA

> Melting point = 1356 K
> Entropy of fusion = 2.29 cal/(mol·K)
> Specific volume = 7 cm³/mol
> Maximum observed supercooling is 236 K

6.3 What is the radius of the critical-sized nucleus of solid copper at 236 K below the melting point using the information in Problem 6.2?

6.4 The temperature at which nuclei of solid water (ice) form homogeneously from undercooled water is −40°C.

(a) What is the critical radius of the solid water nuclei at this temperature?

(b) Why do ponds freeze when the temperature is just a few degrees below the equilibrium freezing point (0°C)?

DATA

Interfacial energy between solid and liquid water is 25 ergs/cm².
Latent heat of fusion of ice is 335 J/g.
Density of ice is 0.92 g/cm³.

6.5 A metal (m) being deposited on a solid oxide surface (s) from the vapor (v) condenses as spherical caps. The contact angle θ is 90°, as indicated in the accompanying diagram.

(a) What is the relationship between the surface energy of the solid surface (γ_{sv}) and the interfacial energy between the metal and the surface (γ_{sm})?

(b) Derive an equation for the critical radius size of a nucleus of the metal as a function of the surface energy of the metal (γ_{mv}) and the volumetric change of Gibbs free energy ($\Delta \underline{G}_v$).

(c) Derive an equation for the Gibbs free energy of the critical nucleus as a function of the surface energy of the metal (γ_{mv}) and the change of Gibbs free energy per unit volume ($\Delta \underline{G}_v$).

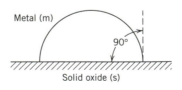

Metal (m)   90°   Solid oxide (s)

6.6 Cylindrical particles of a metal are being grown on the flat surface of the same metal. The height of the particles is constant (a). (This could represent the growth of a new layer of atoms.) Derive an equation for the critical radius of the particles and for the critical free energy change necessary for the nucleation of the particles in terms of the solid–vapor interfacial energy, the entropy of sublimation, and the specific volume of the metal.

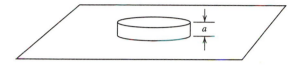

Chapter 7

Reaction Kinetics

The principles discussed in the preceding chapters can now be combined to study the rates of change in various chemical and physical systems. This is the study of reaction kinetics. In this chapter, we consider reactions of three different types: homogeneous reactions in fluids, solid state reactions, and heterogeneous reactions.

In homogeneous reactions, the reactants and products exist in the same phase— for example, two gases reacting to form a third gas. In heterogeneous reactions, the reactants and products exist in different phases. An example is the reaction between gaseous oxygen and solid aluminum to form solid aluminum oxide. Solid state reactions, as the words imply, are reactions in which the reactants and products are in the solid state. The kinetics in solid state reactions differ from the other two classes because of the relatively low mobility of the reactants and the products. An example of this type of transformation is the recrystallization of a cold-worked piece of metal to form relatively strain-free grains. Another is the growth of crystallites in an amorphous polymer, or a glass.

Before studying the individual reaction types, we will review the terminology used in the study of chemical reaction kinetics.

7.1   RATE OF REACTION

Consider the following reaction in which "a" moles of material A react with "b" moles of B to form "c" moles of C and "d" moles of D.

$$aA + bB = cC + dD \tag{7.1}$$

The number of moles of C at any time after the reaction has begun (n_c) can be expressed as follows:

$$n_C = n_C^\circ + c\xi \tag{7.2}$$

where ξ is the extent of the reaction.

In general form, Eqs. 7.1 and 7.2 are written as Eqs. 7.3a and 7.3b:

$$0 = \sum_i \nu_i A_i \tag{7.3a}$$

$$n_i = n_i^\circ + \nu_i \xi \tag{7.3b}$$

where ν_i represents stoichiometric coefficients (negative for reactants, positive for products).

Take as an example the following situation:

3A	+	B	=	C	+	2D	
1		1		0		0	Before reaction (n_i°)
$1 - 3\xi$		$1 - \xi$		ξ		2ξ	After reaction (ξ)

The rate of reaction R is defined as the rate of change of ξ with time.

$$R = \frac{d\xi}{dt} \tag{7.4}$$

The rate of change of the number of moles of species i can be written as follows:

$$\frac{dn_i}{dt} = \nu_i \frac{d\xi}{dt} \tag{7.5}$$

If species involved in a chemical reaction are contained in a volume (which is assumed to be constant), Eq. 7.5 can be written in terms of concentrations:

$$C_i = \frac{n_i}{V}$$

$$\frac{1}{V}\frac{d\xi}{dt} = \frac{1}{\nu_i}\frac{dC_i}{dt} \tag{7.6}$$

7.2   ORDER OF REACTIONS

When reaction rates are determined experimentally, it is often found that the expression for the rate or the extent of reaction can be expressed in the following way[1]:

$$\frac{1}{V}\frac{d\xi}{dt} = kC_A^{\alpha}C_B^{\beta} \quad \text{or} \quad k[A]^{\alpha}[B]^{\beta} \tag{7.7}$$

The exponents of the concentrations, α and β, are called the orders of the reaction. In this case the reaction would be called "of order α" with respect to A, "of order β" with respect to B, or "of order α plus β" overall.

As an example, consider the dissociation of hydrogen iodide into hydrogen and iodine:

$$2HI \rightarrow H_2 + I_2$$

For this reaction it has been found experimentally that

$$-\frac{d[HI]}{dt} = k[HI]^2$$

Because of this relationship, this decomposition is a second-order reaction. It is important to note that the order of a reaction is not necessarily related to the stoichiometric coefficients (a, b, c, d). For example, the reaction of ozone to form oxygen can be written as follows:

$$2O_3 \rightarrow 3O_2$$

If the reaction were an elementary bimolecular reaction, it would be expected to be second order with relation to ozone as follows:

$$-\frac{d[O_3]}{dt} = k\,[O_3]^2$$

Actually, measurements have shown that

$$-\frac{d[O_3]}{dt} = k\,\frac{[O_3]^2}{[O_2]}$$

---

[1]The square bracket notation for concentration—for example, [C] for the concentration of C—will be used interchangeably with the symbol C_C.

This suggests a much more complex reaction path for the formation of oxygen from ozone. One explanation might entail the assumption that ozone breaks down into oxygen and monatomic oxygen in the following reaction:

$$O_3 \underset{k_{-1}}{\overset{k_1}{\rightleftarrows}} O_2 + O$$

Further assume that this monatomic oxygen reacts with ozone to form two diatomic oxygen molecules.

$$O + O_3 \underset{k_2}{\rightarrow} 2O_2$$

If the first reaction proceeds rapidly so that equilibrium is achieved, then the equilibrium constant for the reaction will be:

$$[O] = K \frac{[O_3]}{[O_2]}$$

where $K = \dfrac{k_1}{k_{-1}}$ (see section 7.3)

And for the second reaction we will have

$$-\frac{d[O_3]}{dt} = k_2[O][O_3]$$

Then the overall reaction rate can be expressed as follows:

$$-\frac{d[O_3]}{dt} = k_2[O][O_3] = k_2K \frac{[O_3]^2}{[O_2]}$$

This is one explanation for the form of the experimentally determined rate of the reaction.

The important point made by the example above is that the order of reaction is not necessarily related to the stoichiometric coefficients.

The order of a reaction with respect to a particular component can be determined from experimental data. The data needed are the concentrations of that component as a function of time in the reaction of interest. As an example, consider the reaction of A and B to form another compound. The rate of reaction is expressed by Eq. 7.7:

$$\frac{1}{V} \frac{d\xi}{dt} = k[A]^\alpha [B]^\beta$$

Our task is to determine α, the order of the reaction with respect to A, from data. Let us assume that the data are reliable and accurate enough to determine the slope of the plot of $[A]$ versus time, $d[A]/dt$. From Eq. 7.6,

$$\frac{1}{V}\frac{d\xi}{dt} = -\frac{d[A]}{dt} = k[A]^{\alpha}[B]^{\beta} \tag{7.8}$$

Taking the logarithm of both sides of eq. 7.8,

$$\ln\left(-\frac{d[A]}{dt}\right) = \ln k + \beta \ln [B] + \alpha \ln [A]$$

Assuming that $[B]$, the concentration of B, does not change appreciably during the reaction, the slope of the graph of the logarithm of $(-d[A]/dt)$ versus the logarithm of $[A]$ is α, the order of the reaction with respect to A.

7.3   EQUILIBRIUM CONSTANTS: RELATION TO REACTION RATE CONSTANTS

Consider the reaction of a molecule dissociating to form two ions:

$$MX \rightarrow M^+ + X^- \tag{7.8}$$

The equilibrium constant for this reaction (considering all the species to be present as ideal solutions) would be, in terms of concentrations:

$$K = \frac{[M^+][X^-]}{[MX]} \tag{7.9}$$

If we think of this reaction in kinetic terms, we can define a reaction rate constant in the forward and reverse directions (f^+ and f^-). Assuming first-order kinetics, the rate of change of the concentration of the molecule MX is

$$MX \underset{f^-}{\overset{f^+}{\rightleftarrows}} M^+ + X^- \tag{7.10}$$

$$\frac{d[MX]}{dt} = -f^+[MX] + f^-[M^+][X^-] \tag{7.11}$$

At equilibrium, the rate of change of the concentration of this salt will be zero; hence,

$$\frac{d[MX]}{dt} = 0 \tag{7.12}$$

$$f^+[MX] = f^-[M^+][X^-] \tag{7.13}$$

$$\frac{[M^+][X^-]}{[MX]} = \frac{f^+}{f^-}$$

which implies

$$\frac{f^+}{f^-} = K_{\text{equilibrium}} \tag{7.14}$$

The ratio of the forward and reverse reaction rates is the equilibrium constant. Thus we can conclude that chemical equilibrium is not necessarily a static situation. Rather, we can view it as a balance between forward and reverse reaction rates so that the concentration of the various species remains constant, but the form of the individual chemical reactants and products may change with time. In the solid state, for example, we have the formation and annihilation of Schottky defects. Schottky defects may, through time, be formed and destroyed at various sites in the crystal. But at equilibrium, the product of the concentrations of vacancies on the anion and cation sites is constant, $[V'_M] \times [V^\cdot_X] = K$. At equilibrium, the rate of formation of the defects must equal their rate of annihilation.

7.4 FIRST-ORDER REACTIONS

An example of a first-order reaction is the decomposition of material A into a product (or products). As a simplification, let us first treat a case in which only the forward reaction (from the reactant to the product) is important. This means that the reverse rate, f^-, is negligible in comparison with the forward rate, f^+. Said another way, the equilibrium constant is large; that is, the reaction strongly favors the product or products. Because the reaction is first order, reaction rate and concentrations are related as follows:

$$A = \text{products}$$

from

$$\frac{1}{V}\frac{d\xi}{dt} = \frac{1}{\nu_i}\frac{dC_i}{dt} \quad \text{with} \quad \nu_A = -1$$

$$\frac{1}{V}\frac{d\xi}{dt} = -\frac{d[A]}{dt}$$

If the reaction is first order:

$$\frac{1}{V}\frac{d\xi}{dt} = k[A] \tag{7.15}$$

Thus, the simple differential equation relating concentration and time is

$$-\frac{d[A]}{dt} = k[A] \tag{7.16}$$

Assuming that the concentration of material A at time zero is $[A_0]$, the integrated form of the equation can be derived as follows:

$$\int_{[A_0]}^{[A]} \frac{d[A]}{[A]} = -k \int_0^t dt$$

$$\ln \frac{[A]}{[A_0]} = -kt \tag{7.17}$$

$$[A] = [A_0] \exp(-kt)$$

A good example of this kind of reaction is the decay of a radioactive species. In this case the relationship between the number of moles of the radioactive species present at any time (N) and the amount originally present at time zero (N_0) is:

$$N = N_0 \exp(-\lambda t) \tag{7.18}$$

where λ is called the decay constant.

By taking the natural logarithm of both sides of Eq. 7.18, we can see that a graph of the natural log of the number of moles of the species present at any time versus time will be a straight line with the slope negative λ.

$$\ln N = \ln N_0 - \lambda t$$

In the case of radioactive species, this decay constant is often expressed in terms of the half-life of the species (τ), the time in which the concentration is reduced to half the original concentration. In this case $N/N_0 = \frac{1}{2}$; thus:

$$\frac{N}{N_0} = \frac{1}{2} = \exp(-\lambda\tau)$$

$$2 = \exp(\lambda\tau) \qquad \text{or} \qquad \lambda = \frac{\ln 2}{\tau}$$

Substituting in Eq. 7.18:

$$N = N_0 2^{-t/\tau} \quad \text{or} \quad N = N_0 \left(\frac{1}{2}\right)^{t/\tau} \tag{7.19}$$

An application of Eq. 7.19 is in the carbon dating of archaeological samples. Carbon-14 is a radioactive isotope of carbon with a half-life of 5760 years. Cosmic radiation in the upper atmosphere synthesizes carbon-14, which balances the loss due to radioactive decay. Living matter exchanges carbon with the atmosphere and maintains a level of carbon-14 that produces 15.3 disintegrations per minute per gram of carbon contained. Dead organisms no longer exchange carbon with atmospheric CO_2, and the amount of carbon-14 in dead material decreases with time as a result of radioactive decay. As an example, let us calculate the age of an archaeological sample that undergoes 10 disintegrations per minute per gram of carbon in the sample. Because we are dealing with a first-order reaction, the rate of disintegration is proportional to the amount of carbon-14 present. Hence:

$$\frac{N}{N_0} = \frac{10}{15.3} = 0.65$$

From Eq. 7.18,

$$\ln \frac{N}{N_0} = -\lambda t = \ln 0.65 = -0.43$$

From the definition of the half-life,

$$\lambda = \ln \frac{2}{\tau} = \ln \frac{2}{5760} = 1.23 \times 10^{-4} \text{ year}^{-1}$$

Hence, the age of the sample is about

$$t = \frac{0.43}{1.23 \times 10^{-4}} = 3530 \text{ years,}$$

better stated as 3500 years, considering the accuracy possible in these measurements.

7.5 FIRST-ORDER REACTIONS WITH FORWARD AND REVERSE RATES

Section 7.4 dealt with the type of first-order reaction, such as radioactive decay, in which the forward rate dominates. It is hard to imagine that process of radioactive decay reversing itself spontaneously.

In this section we consider a first-order reaction in which both the forward and reverse reactions must be considered. We will demonstrate that the rate of a reaction (the reaction rate constant) depends on the driving force (the decrease in Gibbs free energy) when the driving force is small compared to RT, as in the case of nucleation or diffusion. The reaction rate will be independent of the magnitude of the driving force when the driving force is large compared to RT, as in radioactive decay or oxidation reactions with large values of ΔG.

Consider the reaction below.

$$A \underset{f^-}{\overset{f^+}{\rightleftarrows}} B \tag{7.20}$$

If we assume first-order kinetics for both the forward and reverse reactions, then:

$$-\frac{d[A]}{dt} = -[\dot{A}] = f^+[A] - f^-[B] \tag{7.21}$$

At equilibrium, the rate of change of the concentration of A, $[\dot{A}]$, is zero, hence:

$$\frac{f^+}{f^-} = \frac{[B]_e}{[A]_e} = K = \exp\left(-\frac{\Delta G^\circ}{RT}\right) = \exp\left(-\frac{\mu_B^\circ - \mu_A^\circ}{RT}\right)$$

where $[A]_e$ is the equilibrium concentration of A.

Substituting in Eq. 7.21 yields

$$-[\dot{A}] = f^+ \left([A] - \frac{[A]_e}{[B]_e}[B]\right) \tag{7.22a}$$

$$-[\dot{A}] = f^+ \left(1 - \frac{[A]_e}{[B]_e}\frac{[B]}{[A]}\right)[A]$$

If we want to use the simple form of the kinetic equation derived in Section 7.4 (Eq. 7.17), we can substitute as follows:

$$f = f^+ \left(1 + \frac{[A]_e}{[B]_e}\frac{[B]}{[A]}\right) \tag{7.22b}$$

To put this expression in terms of the Gibbs free energy, or chemical potential change involved in the reaction, we can make the following substitutions:

$$\mu_A = \mu_A^\circ + RT \ln a_A$$

Assuming ideal solutions,

$$a_A = x_A \quad \text{and} \quad x_A = \frac{N_A}{N_A + N_B}$$

The concentration of A, $[A] = N_A/V$, where N_A is the number of moles of A and V is the volume. Thus:

$$\mu_A = \mu_A^\circ + RT \ln \left\{ [A] \left(\frac{V}{N_A + N_B} \right) \right\}$$

The term $V/(N_A + N_B)$ is the molar volume, $\underline{V}$. Setting the molar volumes of A and B each equal to the overall molar volume,

$$\mu_A = \mu_A^\circ + RT_{\ln} \{ [A]\underline{V} \}$$

or

$$[A] = \frac{1}{\underline{V}} \exp \left(\frac{\mu_A - \mu_A^\circ}{RT} \right)$$

The same is true for component B.
Then:

$$\frac{[B]}{[A]} = \exp \left(\frac{\mu_B - \mu_A}{RT} \right) \exp \left(\frac{\mu_A^\circ - \mu_B^\circ}{RT} \right) = \exp \left(\frac{\mu_B - \mu_A}{RT} \right) \exp \left(-\frac{\Delta G^\circ}{RT} \right)$$

But,

$$\frac{[B]_e}{[A]_e} = K = \exp \left(-\frac{\Delta G^\circ}{RT} \right)$$

Hence,

$$\frac{[B]}{[A]} = \exp \left(\frac{\mu_B - \mu_A}{RT} \right) \frac{[B]_e}{[A]_e}$$

Substituting in Eq. 7.22b, we have

$$-[\dot{A}] = f^+ \left\{ 1 - \exp \left(\frac{\mu_B - \mu_A}{RT} \right) \right\} [A]$$

Thus the first-order rate constant is

$$f = f^+ \left\{ 1 - \exp \left(\frac{\mu_B - \mu_A}{RT} \right) \right\} \tag{7.23}$$

This expression is sometimes written in terms of the driving force for the reaction ΔG_r, or $\mu_A - \mu_B$, which is the negative of the ΔG of reaction as normally written.

$$f = f^+ \left\{ 1 - \exp\left(-\frac{\Delta G_r}{RT} \right) \right\} \qquad (7.24)$$

When the driving force for the reaction is large (i.e., ΔG is large as compared to RT), the exponential term in Eq. 7.24 becomes negligible compared to 1 and the overall reaction rate is simply the forward rate as in Section 7.4.

$$f = f^+ \left\{ 1 - \exp\left(-\frac{\Delta G_r}{RT} \right) \right\} = f^+ \qquad (7.25a)$$

One such case has already been discussed, that of radioactive decay. Another illustration is the oxidation of silicon to form silicon dioxide. At 900°C (1173 K), the $\Delta G°$ for the reaction is about $-700,000$ J/mol. The term $\Delta G°/RT$ is about 72, and $\exp(-\Delta G°/RT)$ is about 6×10^{-32}. Thus the frequency of the reverse reaction is negligible, and the overall reaction rate is simply equal to the forward reaction rate.

By contrast, when ΔG_r is small compared to RT, Eq. 7.24 becomes (noting that $\exp x = 1 - x$ when x is small):

$$f = f^+ \left\{ 1 - 1 + \frac{\Delta G_r}{RT} \right\}$$

$$f = f^+ \left\{ \frac{\Delta G_r}{RT} \right\} \qquad (7.25b)$$

In many condensed state reactions ΔG_r is indeed small compared to RT. Grain growth in a solid is one example. In a solid with an average grain size of 0.1 mm, there is about 300 cm^2 of grain boundary per cubic centimeter of material. The driving force for grain growth is the reduction of total interfacial energy. Assuming a specific interfacial energy of about 300 ergs/cm^2 gives a total grain boundary energy of the order of 10^5 ergs per cubic centimeter of material, or about 0.1 J per mole of material, assuming a molar volume of about 10 cm^3/mol. The term $\Delta G_r/RT$ at a temperature of about 900 K is about 3×10^{-5}, a value small compared to 1. This means that the reaction rate for grain growth takes the form of Eq. 7.25b. The reaction rate depends on the magnitude of the driving force in the case at hand. There is no universal rate that characterizes grain growth in a particular material. The recrystallization of cold-worked material is another example. The degree of cold work (or amount of energy stored in the lattice distortion) will influence the reaction rate as the material recrystallizes to form stress-free grains.

As another example, consider the process of diffusion as a reaction. In this case, we think of the reaction as the jump of atoms from one lattice position to another, a movement through a distance λ. The velocity of the atom is

$$v = f\lambda \tag{7.26a}$$

where f is the "reaction rate."

The flux of diffusing atoms is

$$J = vC = f\lambda C \tag{7.26b}$$

where C is the concentration of the diffusing species. Expressing the reaction rate according to Eq. 7.25b because the term $\Delta G_r/RT$ is small, we write

$$f = f^+ \left\{ \frac{\Delta G_r}{RT} \right\}$$

Substituting for the chemical potential and assuming ideal solutions:

$$\Delta G_r = -\left(\frac{\partial \mu}{\partial x}\right)_T \Delta x = -\left(\frac{\partial \mu}{\partial x}\right)_T \lambda = -\left(\frac{\partial(RT \ln C)}{\partial x}\right)_T \lambda$$

At constant T,

$$\frac{\Delta G_r}{RT} = -\frac{dC}{dx}\left(\frac{1}{C}\right)\lambda \quad \text{or} \quad f = -f^+\lambda \frac{1}{C}\frac{dC}{dx}$$

$$J = f\lambda C = -f^+\lambda^2 \frac{dC}{dx}$$

Thus the reaction rate approach yields a relationship that is parallel to Fick's first law, where the diffusion coefficient D is

$$D = f^+\lambda^2 \tag{7.27}$$

7.6   HIGHER ORDER REACTIONS

The mathematical representation of the kinetics of higher order reactions is a straightforward exercise in calculus. For a second-order reaction of the form

$$2A = \text{products} \tag{7.28}$$

the differential equation relating concentration and time is:

$$-\frac{d[A]}{dt} = k[A]^2 \qquad \textbf{(7.29a)}$$

In integrated form this yields:

$$\frac{1}{[A]} - \frac{1}{[A_0]} = kt \qquad \textbf{(7.29b)}$$

Consider a second-order reaction involving reactants A and B:

$$A + B = products$$

The reaction rate equation is:

$$\frac{d[A]}{dt} = -k[A][B] \qquad \textbf{(7.30)}$$

The integrated form of the Eq. 7.30 depends on the relationship of the initial concentrations of A and B, $[A_0]$ and $[B_0]$. If $[A_0] = [B_0]$, then:

$$\frac{1}{[A]} = \frac{1}{[A_0]} + kt \qquad \text{or} \qquad \frac{1}{[B]} = \frac{1}{[B_0]} + kt \qquad \textbf{(7.31)}$$

If the initial concentrations of A and B, $[A_0]$ and $[B_0]$, are different, then:

$$\ln\left(\frac{[A]}{[A_0]}\right) - \ln\left(\frac{[B]}{[B_0]}\right) = -k([B_0] - [A_0])t \qquad \textbf{(7.32a)}$$

An interesting special case arises when one of the reactants, B for example, is present in very much larger concentration than A. During the reaction, the concentration of B will not change much, i.e., $[B] = [B_0]$ or $[B]/[B_0] = 1$. Eq. 7.32a then becomes:

$$\ln\frac{[A]}{[A_0]} = -(k[B_0])t \qquad \textbf{(7.32b)}$$

The reaction, thus, becomes essentially first-order with respect to A. The rate constant is $k[B_0]$. (See text following Eq. 7.8.)

For third-order reactions of the kind:

$$3A = products \qquad \textbf{(7.33)}$$

The corresponding equations are:

$$-\frac{d[A]}{dt} = k[A]^3 \tag{7.34a}$$

$$\frac{1}{[A]^2} - \frac{1}{[A_0]^2} = kt \tag{7.34b}$$

7.6   REACTIONS IN SERIES

In this section we consider a situation in which two reactions may take place in series. Material A reacts (or decomposes) into material B, which in turn forms material C. Diagrammatically this can be shown as

Through the mathematical manipulations to be shown, we will illustrate that the overall rate of reaction A to C will be governed by the relative rates of reaction f_1 and f_2. The slower of the two will control the overall reaction.

For simplicity, assume that the reactions are all first order, and that at zero time the concentration of A is $[A_0]$ and the concentrations of B and C are zero. The differential equations that govern the concentrations of the various species are as follows:

$$-\frac{d[A]}{dt} = f_1[A] \tag{7.35a}$$

$$-\frac{d[B]}{dt} = -f_1[A] + f_2[B] \tag{7.36a}$$

$$\frac{d[C]}{dt} = f_2[B] \tag{7.37a}$$

The integrated forms of these equations are:

$$[A] = [A_0] \exp(-f_1 t) \tag{7.35b}$$

$$[B] = \frac{f_1}{f_2 - f_1} [A_0] \exp(-f_2 t)[\exp(f_2 - f_1) - 1] \tag{7.36b}$$

$$[C] = [A_0] \left[1 - \frac{f_2}{f_2 - f_1} \exp(-f_1 t) + \frac{f_1}{f_2 - f_1} \exp(-f_2 t) \right] \tag{7.37b}$$

In the first case, consider the situation in which the reaction rate for A to B (f_1) is very much slower than f_2, the reaction between B and C. In this case the concentration of material C becomes:

$$[C] = [A_0][1 - \exp(-f_1 t)] \quad \text{or} \quad \frac{d[C]}{dt} = f_1([A_0] - [C]) \quad \textbf{(7.38)}$$

This illustrates that reaction from A to C is controlled by the reaction rate constant f_1, the reaction from A to B.

In the second case, assume that the reaction rate constant from B to C (f_2) is slower than the one from A to B (f_1).

$$[C] = [A_0][1 - \exp(-f_2 t)] \quad \text{or} \quad \frac{d[C]}{dt} = f_2([A_0] - [C]) \quad \textbf{(7.39)}$$

The foregoing is an example of what is often called the *bottleneck principle*. This implies that in a series of reactions, the overall rate of reaction is controlled by the bottleneck rate, that is, the slowest rate in the series.

7.8 TEMPERATURE DEPENDENCE OF REACTION RATE

One of the factors that can be used to control the rate of a reaction is the temperature at which the reaction takes place. Usually higher temperatures mean faster reaction rates.

The seminal work on the subject was that of Arrhenius in the latter part of the nineteenth century. He observed that the rate of change of the equilibrium constant for a reaction with temperature could be expressed as follows:

$$\frac{d(\ln Ka)}{dT} = \frac{\Delta H}{RT^2} \quad \textbf{(7.40)}$$

But, as we have shown, the equilibrium constant can really be thought of as the ratio of two reaction rate constants, the forward and the reverse reaction rates ($K_a = f^+/f^-$). It was thus a reasonable assumption that these individual reaction rates should follow the same basic mathematical form:

$$\frac{d(\ln f^+)}{dT} = \frac{E^*}{RT^2} \quad \text{or} \quad d(\ln f^+) = -\frac{E^*}{R} d\left(\frac{1}{T}\right)$$

where E^* is called an activation energy. Integrating the equation yields:

$$\ln f^+ = A \exp\left(-\frac{E^*}{RT}\right) \quad \textbf{(7.41)}$$

From Eq. 7.41 it is apparent that a graph of the natural logarithm of the reaction rate constant as a function of inverse absolute temperature should be a straight line with a negative slope of the activation energy (E) divided by R, the universal gas constant. The preexponential factor A is sometimes called the frequency factor.

One way of looking at this is to consider a chemical potential barrier between the reactant A and its decomposition products (Figure 7.1). We can think of a pseudo-equilibrium between the material A and A*, where A* is an activated complex on the path between A and its decomposition products.

The equilibrium constant relating the concentration of this activated complex to the concentration of the reactant is:

$$K = \frac{[A^*]}{[A]} = \exp\left(-\frac{\Delta G^*}{RT}\right) \qquad \text{(7.42a)}$$

where ΔG^* is the Gibbs free energy of formation of this activated complex from the reactant A.

The forward reaction rate is determined by the concentration of the activated species, $[A^*]$, and the decomposition frequency, $f°$.

$$f^+ = [A^*]f°$$

This yields an equation of the form:

$$f^+ = f° \exp\left(-\frac{\Delta G^*}{RT}\right) = f° \exp\left(\frac{\Delta S^*}{R}\right) \exp\left(-\frac{\Delta H^*}{RT}\right) \qquad \text{(7.42b)}$$

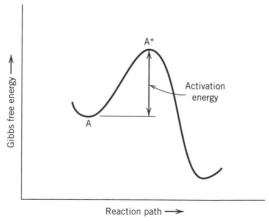

Figure 7.1 Schematic diagram of activation energy along reaction path.

The decomposition rate of the activated complex is governed by its vibration pattern. By a consideration of the energy required to disintegrate the complex it can be shown that the decomposition frequency, $f°$, is kT/h, where h is the Planck constant and k is the Boltzmann constant. Based on this approach, the reaction rate constant, f^+, is:

$$f^+ = \exp\left(\frac{\Delta S^*}{R}\right) \frac{kT}{h} \exp\left(-\frac{\Delta H^*}{RT}\right) \qquad \textbf{(7.42c)}$$

The rate constant, using the activation energy approach, is then dependent on temperature as follows[2]:

$$f^+ \propto T \exp\left(-\frac{\Delta H^*}{RT}\right) \qquad \text{or} \qquad T \exp\left(-\frac{E^*}{RT}\right) \qquad \textbf{(7.42d)}$$

One may also derive the temperature dependence of the reaction rate constant in gas reactions by postulating that molecules that undergo a collision with energy greater than a specified level, E^*, will react. Consider a reaction in which two molecules of A form another compound. The collision rate of two molecules of A per unit volume, N_{AA}/t, is:

$$\frac{N_{AA}}{t} = \sigma_A^2 \left(\frac{4\pi kT}{M_A}\right)^{1/2} [A]^2 \qquad \textbf{(7.43a)}$$

where σ_A is the diameter of the molecule, and M_A is the mass of the A molecule.

In terms of temperature, this equation is of the form:

$$\frac{N_{AA}}{t} \propto T^{1/2}[A]^2 \qquad \textbf{(7.43b)}$$

If we postulate that only collisions with energy greater than E^* will result in a reaction, then the reaction rate constant will be of the form:

$$f^+ \propto T^{1/2} \exp\left(-\frac{E^*}{RT}\right) \qquad \textbf{(7.43c)}$$

We thus have three forms of the temperature dependence of the reaction rate constant, f^+. We can express f^+ as:

$$f^+ \propto T^m \exp\left(-\frac{E^*}{RT}\right) \qquad \textbf{(7.44a)}$$

---

[2]The terms ΔH^* and E^* will be used interchangeably in this section.

The value of the factor m can be between zero and one. Taking the logarithm of Eq. 7.44a and differentiating with respect to temperature:

$$\ln f^+ = m \ln T - \frac{E^*}{RT} \tag{7.44b}$$

$$\frac{d \ln f^+}{dT} = \frac{m}{T} + \frac{E^*}{RT^2} = \frac{mRT + E^*}{RT^2} \tag{7.44c}$$

If we note that mRT is usually much less than E^*, it becomes apparent that the temperature dependence of the reaction rate constant can well be approximated by the Arrhenius form of Eq. 7.41.

7.9   HETEROGENEOUS REACTIONS

Most studies concerned with chemical reaction rates deal with homogeneous chemical reactions, that is, reactions occurring within a single fluid phase. However, many of the reactions of interest to materials scientists and engineers are heterogeneous; that is, reactants are initially present in different phases and therefore have to react at a phase boundary. These heterogeneous reactions take place by way of a series of consecutive steps. Consider the simple reaction of a gas dissolving in a metal— for example, hydrogen dissolving in aluminum (Figure 7.2a). The overall reaction (Figure 7.2b) may be written as follows:

$$H_2 \text{ (gas)} = 2\underline{H} \text{ (in solution)}$$

This expression represents the stoichiometry of the equilibrium involved. It does not, however, express the mechanism of the reaction, which consists of a series of consecutive steps.

1. The transport of hydrogen molecules by diffusion (or convection) in the gas phase to the gas–metal phase boundary.
2. The adsorption of hydrogen molecules on the surface of the aluminum.
3. The decomposition of the adsorbed molecules into adsorbed hydrogen atoms.
4. The solution of adsorbed hydrogen atoms into the aluminum at the gas–metal phase boundary.
5. The transport of the dissolved hydrogen atoms away from the phase boundary.

In principle, all these steps can be regarded as limiting the rate of the overall reactions. They act as resistances in series. In practice, however, it is usually one step that is slow enough (the bottleneck) to effectively determine the overall rate of reaction. (This was illustrated in Section 7.7.) In the example cited, possible rate-limiting steps include the transport of hydrogen to the gas–metal interface or away from the interface into the metal, and the surface reaction itself.

Because each of these consecutive reactions has its own activation energy, hence its own temperature dependence, the rate-limiting reaction may vary as the temperature is varied, or as other physical parameters are varied.

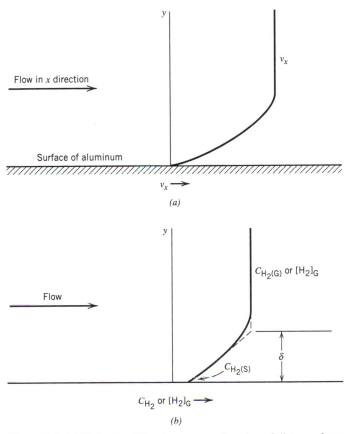

Figure 7.2 (*a*) Velocity of flowing gas as a function of distance from the surface. (*b*) Concentration of hydrogen as a function of distance from the surface.

As an example, consider the situation of diffusion to the gas–metal interface. Assume, for the time being, that the hydrogen is contained in a carrier gas such as argon. The diffusion of hydrogen to the surface may well be controlled by the diffusion of the hydrogen through a boundary layer (Figure 7.2) which is in turn dependent on the flow rate of the gases over the metallic surface as follows:

$$J_{H_2} = -D_{H_2} \frac{d[H_2]}{dx}$$

$$J_{H_2} = -D_{H_2} \left\{ \frac{[H_2]_S - [H_2]_G}{\delta} \right\}$$

$$J_{H_2} = -\frac{D_{H_2}}{\delta} \{[H_2]_S - [H_2]_G\} \tag{7.45}$$

In this case the temperature dependence of the reaction rate would not be of the Arrhenius form (Eq. 7.41). The temperature dependence is determined by the rate-limiting step, the diffusion through the boundary layer. In that case the rate depends on $T^{3/2}$, as discussed in Chapter 2 (Section 2.14).

The oxidation of silicon by an oxygen-containing gas provides an interesting example (Ref. 1). Figure 7.3 illustrates the oxygen concentration and chemical potentials after some silicon dioxide has been formed. The rates of oxygen movement or consumption in the three phases present are:

$$F_1 = h(\mu_G - \mu_S) \qquad \text{[in gas]}$$

$$F_2 = D\frac{\mu_S - \mu_I}{x} \qquad \text{[oxide]}$$

$$F_3 = k\mu_I \qquad \text{[at metal-oxide interface]}$$

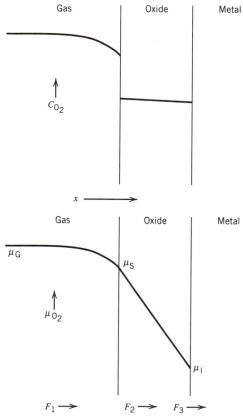

Figure 7.3 Concentration of oxygen (C_{O_2}) and chemical potential of oxygen (μ_{O_2}) in the oxidation of silicon metal.

At steady state, $F_1 = F_2 = F_3$. We have established that the slowest step will limit the rate of reaction. In this case, it is known that F_1 is fast compared to F_2 and F_3. This simplifies the situation, as shown in Figure 7.4.

Because $F_2 = F_3$,

$$\frac{D(\mu_S - \mu_I)}{x} = k\mu_I$$

and $\mu_s = \mu_G$, and μ_I is

$$\mu_I = \frac{D\mu_G}{kx + D} = \frac{\mu_G}{kx/D + 1}$$

An expression for F_2 is then:

$$F_2 = D\frac{(\mu_G - \mu_I)}{x} = \frac{D}{x}\left[\mu_G - \frac{\mu_G}{1 + kx/D}\right]$$

$$F_2 = \frac{Dk\mu_G}{D + kx}$$

But F_2 is the rate of growth of the silicon dioxide layer

$$F_2 = \frac{dx}{dt}N$$

where N is a constant related to the volume change from silicon to silicon dioxide.

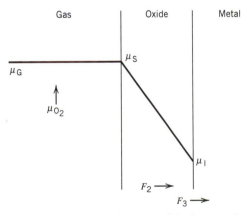

Figure 7.4 The chemical potential of oxygen in the oxidation of silicon metal with very fast diffusion in the gas phase relative to transport in the oxide or reaction at the oxide–metal interface.

Then,

$$\frac{dx}{dt} = \frac{F_2}{N} = \frac{Dk\mu_G}{N(D + kx)}$$

$$(D + kx)dx = \frac{Dk\mu_G}{N} dt$$

Integrating, we have

$$\int_0^x (D + kx) \, dx = \frac{Dk\mu_G}{N} \int_0^t dt$$

$$Dx + \frac{k}{2} x^2 = \frac{Dk\mu_G}{N} t$$

$$x^2 + \frac{2D}{k} x - \frac{2D\mu_G}{N} t = 0$$

This equation is of the form:

$$x^2 + Bx - Ct = 0$$

where $B = 2D/k$ and $C = \dfrac{2D\mu_G}{N}$

$$x = \frac{-B + (B^2 + 4Ct)^{1/2}}{2}$$

$$x = \frac{B}{2}\left[\left(1 + \frac{t}{B^2/4C}\right)^{1/2} - 1\right]$$

Now consider two cases. In case I, $t \ll B^2/4C$. Then, $(1 + \varepsilon)^{1/2} = 1 + \varepsilon/2$

$$x = \frac{C}{B} t = \frac{k\mu_G}{N} t$$

and the thickness x is linear with time; that is, the reaction rate is constant.
   In case II, where $t \gg B^2/4C$,

$$x = \frac{B}{2}\left[\frac{t}{B^2/4C}\right]^{1/2} = (Ct)^{1/2}$$

$$x^2 = Ct = \frac{2D\mu_G}{N} t$$

The thickness x is parabolic with time; that is, it is a function of the square root of time. This is illustrated in Figure 7.5.

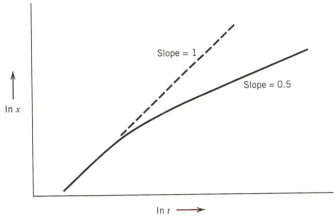

Figure 7.5 Thickness of oxide (x) versus time for the oxidation of silicon metal.

In the preceding example, the rate-limiting step changed during the course of the reaction, and the rate changed accordingly.

7.10   SOLID STATE TRANSFORMATION KINETICS

Many of the reactions of interest to materials scientists involve transformations in the solid state, reactions such as recrystallization of a cold-worked material, the precipitation of a crystalline polymer from an amorphous phase, or simply the growth of an equilibrium phase from a nonequilibrium structure, the driving force for which is brought about by cooling from one temperature to another. Consider the initial phase to be α and the resulting phase to be β; the reaction is thus written:

$$\alpha \rightarrow \beta$$

The total volume of the sample is the sum of the volumes of α and β:

$$V = V^\alpha + V^\beta \tag{7.46}$$

The fraction transformed is simply:

$$F = \frac{V^\beta}{V} \tag{7.47}$$

Assume that the process of transforming α to β is controlled by nucleation and growth, that is, the nucleation of phase β within α and then the rate of growth of β.

Consider that:

$\dot{N}$ = nucleation rate per unit volume

$\dot{G}$ = growth rate in one direction = dr/dt (assuming spherical form of β)

Referring to Figure 7.6, consider the time line from zero to a time, t. We will consider another measure of time (τ), which starts when a nucleus is formed. The number of nuclei formed in the differential time $d\tau$ is equal to

$$\dot{N}V^{\alpha}d\tau$$

Assuming that the particles nucleated in this time $d\tau$ grow as spheres, the radius of the particles formed during $d\tau$, after they have grown to time t, is:

$$\int_0^r dr = \int_\tau^t \dot{G}\, d\tau$$

$$r = \dot{G}(t - \tau) \tag{7.48}$$

The volume of the particle nucleated during $d\tau$ at time t is:

$$dV^{\beta} = \tfrac{4}{3}\pi \dot{G}^3(t - \tau)^3(\dot{N}V^{\alpha})d\tau$$

$$dV^{\beta} = \tfrac{4}{3}\pi \dot{G}^3\dot{N}(V - V^{\beta})(t - \tau)^3 d\tau \tag{7.49}$$

Early in the transformation, when V^{β} is small, V^{β} can be considered negligible with respect to V. In this case, the fraction transformed may be calculated as follows:

$$\int_0^{V^{\beta}} dV^{\beta} = \int_0^t \tfrac{4}{3}\pi \dot{G}^3\dot{N}V(t - \tau)^3 d\tau$$

$$V^{\beta} = V\frac{\pi}{3}\dot{G}^3\dot{N}t^4$$

$$F = \frac{V^{\beta}}{V} = \frac{\pi}{3}\dot{G}^3\dot{N}t^4 \tag{7.50}$$

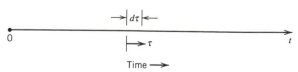

Figure 7.6 Definitions of t and τ for derivation of solid state transformation equations.

To treat the regime beyond the early transformation, we adopt what is called the *extended volume concept*. In this case we separate the nucleation and growth rates from geometrical considerations such as impingement. The extended volume (V_e) is the volume that would have been formed if the entire volume had participated in nucleation and growth, even that portion transformed (V^β). In this case,

$$dV_e^\beta = V \tfrac{4}{3}\pi \dot{G}^3 \dot{N}(t - \tau)^3 d\tau$$

$$V_e^\beta = \tfrac{4}{3}\pi V \int_0^t \dot{G}^3 \dot{N}(t - \tau)^3 d\tau \tag{7.51}$$

But the total volume is equal to the sum of the volumes of α and β:

$$V = V^\alpha + V^\beta$$

$$\frac{V^\alpha}{V} = 1 - \frac{V^\beta}{V} = 1 - F$$

where $F = V^\beta/V$.

The amount of β formed, dV^β, is the fraction of α times dV_e^β

$$dV^\beta = \left(1 - \frac{V^\beta}{V}\right) dV_e^\beta \tag{7.52}$$

Integrating Eq. 7.52,

$$V_e^\beta = -V \ln\left(1 - \frac{V^\beta}{V}\right) = -V \ln(1 - F) \tag{7.53}$$

Combining Eqs. 7.53 and 7.51 yields:

$$-\ln(1 - F) = \tfrac{4}{3}\pi \int_0^t \dot{G}^3 \dot{N}(t - \tau)^3 d\tau$$

If $\dot{G}$ and $\dot{N}$ are constant,

$$-\ln(1 - F) = \tfrac{4}{3}\pi \dot{G}^3 \dot{N} \int_0^t (t - \tau)^3 d\tau = \frac{\pi}{3} \dot{G}^3 \dot{N} t^4$$

$$F = 1 - \exp\left(-\frac{\pi}{3} \dot{G}^3 \dot{N} t^4\right) \tag{7.54}$$

The resulting equation relating the fraction transformed to nucleation rate, growth rate, and time is called the *Johnson–Mehl equation*.

A similar treatment of the subject is given by Avrami. In general he expresses the fraction transformed as

$$F = 1 - \exp(-kt^n) \tag{7.55}$$

where n is called "the Avrami n."

The variation of n from 4 (in the Johnson–Mehl equation) can occur for a number of reasons. In some solid state reactions, the nucleation rate is a decaying function of time. In that case the Avrami n would be 4 early in the reaction, but decreasing to 3 as the nucleation decreases as a function of time, and the transformation is governed strictly by the growth rate. In general, for three-dimensional solids, the Avrami n is between 3 and 4.

In the case of growth of a phase in two dimensions such as in a sheet or a film, the Avrami n is between 2 and 3. In the case of wire, a one-dimensional solid, the Avrami n is between 1 and 2.

To determine the value of the Avrami n from Eq. 7.55, the following mathematical manipulation is performed:

$$F = 1 - \exp(-kt^n)$$
$$1 - F = \exp(-kt^n)$$
$$\ln(1 - F) = -kt^n$$
$$\ln \ln(1 - F) = \ln k - n \ln t$$

Thus the Avrami n is the slope of the plot of the logarithm of the logarithm of $(1 - F)$ versus the negative of the logarithm of t.

REFERENCES

1. Ohring, M., *The Material Science of Thin Films,* Academic Press, San Diego, CA, 1992, Section 8.7.

PROBLEMS

7.1 In a diffusion experiment, radioactive copper (^{64}Cu) is plated on one end of a sample and then "diffused" by heating in a furnace for a specified period of time. The concentration of copper-64 as a function of distance is determined by taking thin slices of the sample and counting the rate of isotope decay in them.

The last slice is analyzed 4 hours after the first. What correction factor must be applied to the last reading to make it comparable to the first?

The half-life of ^{64}Cu is 12.8 hours.

7.2 How many disintegrations occur per minute in a sample containing 1 gram of ^{235}U? The half-life of ^{235}U is 7.04×10^8 years.

7.3 The decay rate of the isotope carbon-14 (^{14}C) is often used to establish the date on which carbon-containing matter died. In the upper atmosphere, cosmic radiation synthesizes ^{14}C. This process balances the loss of ^{14}C through radioactive decay. Living matter, which exchanges carbon with atmospheric carbon dioxide and maintains its ^{14}C level, produces 15.3 disintegrations per minute per gram of carbon it contains. Dead organisms no longer exchange carbon with the atmosphere, and the ^{14}C content decreases with time because of radioactive decay.

  A 0.5 g sample of a plant from an excavation shows 3.5 disintegrations per minute from its ^{14}C. How long ago did the plant die?

7.4 A well-known generalization (rule of thumb) concerning biochemical reactions near room temperature (300 K) is that their rates double for every 10 K rise in temperature.

  (a) What is the activation energy for the reactions implied by this rule?
  (b) In an experiment on one of these reactions conducted near room temperature, the rate constant (k) is to be determined to an accuracy of $\pm 1\%$. How accurately must the temperature be controlled to achieve at least this level of accuracy in k? State answer in $\pm$ degrees kelvin.

7.5 The following statement appears on the side of a milk carton:

  The contents will spoil in 8 days if stored at 5°C.
  The contents will spoil in 12 hours if stored at 30°C.

  If the mechanism and definition of spoilage remain constant in this temperature range, how long will the milk remain unspoiled at 20°C?

7.6 An amorphous polymer fiber is being treated to form polymer crystals. You may assume that the fiber is thin enough to be treated as a one-dimensional solid.

  (a) Write an equation for the functional relationship between the fraction of the amorphous polymer that has crystallized (F) and time (t). Assume that the nucleation of crystals is homogeneous and that the rate of nucleation is constant. Assume also that the rate of crystal growth is constant.
  (b) After 1 hour of the treatment, 10% of the polymer fiber is crystalline. How long will it take for 50% of the fiber to become crystalline?
  (c) Will your answer to part a be changed if the nucleation is *heterogeneous,* and all the nuclei become active at the beginning of the transformation ($t = 0$)? If so, how?

7.7 For simple decomposition reactions (i.e., those with no reverse reaction) the "half-life" concept yields a convenient equation for the amount of reactant remaining as a function of time:

$$\frac{N}{N_0} = \left(\frac{1}{2}\right)^{t/\tau}$$

where N = amount remaining after time t
$\quad N_0$ = initial amount of reactant
$\quad \tau$ = half-life

Is this equation applicable to solid state reactions whose progress can be described by the Johnson–Mehl equation?

$$1 - F = \exp\left[-\frac{\pi}{3} \dot{G}^3 \dot{N} t^4 \right]$$

where F = fraction transformed.

A simple yes or no is insufficient; demonstrate the basis for your answer.

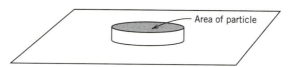
Area of particle

7.8 Using an approach similar to the one used to derive the Johnson–Mehl or Avrami equation, derive a relationship between the fraction of the area covered by the particles, $F = A_c/A$, and the growth rate $(\dot{G})$, the nucleation rate $(\dot{N})$, and time.

$\quad A_c$ = area covered by particles
$\quad A_u$ = area uncovered by particles
$\quad A$ = total area
$\quad F \equiv A_c/A$

7.9 Metallic silicon is being oxidized in an atmosphere of pure oxygen at 1200 K. The thickness of the oxide x (in micrometers) as a function of time t (in seconds) is given by the equation:

$$x^2 + Ax = Bt$$

(a) What is the rate of change of oxide thickness when the oxide thickness (x) is 10 μm? (See data below.)

(b) When the oxide thickness is 10 μm, is the oxidation reaction controlled by diffusion through the oxide film, or by the reaction at the silicon–silicon dioxide surface? Justify your answer.

(c) If the oxygen pressure in the reaction chamber is lowered, do you expect the oxidation rate to change (when $x = 10$ μm)? Why?

DATA

At 1200°C, an oxygen pressure of 600 torr,

$$A = 0.04 \text{ μm and } B = 0.045 \text{ μm}^2/\text{s}$$

At 1200°C, the pressure of oxygen in equilibrium with Si and SiO_2 is approximately 10^{-22} atm.

7.10 Atoms (isotopes) A, B, and C are part of a radioactive decay sequence.

$$A \to B \to C$$

The half-life of A (relative to its decay to B) is one year. The half-life of B (relative to its decay to C) is one hour.

(a) What fraction of the original A is left after one month?
(b) If the original amount of A was 10^{-6} mol, how much C (moles) will be formed after one month?

7.11 During "calcination" of limestone, calcium carbonate ($CaCO_3$) is decomposed into lime (CaO) and carbon dioxide. The rate of calcination is controlled by the rate of mass transport of CO_2 from the surface of the solid to the surrounding atmosphere.

Assume that the carbon dioxide diffuses through a stagnant film of air 1 mm thick and that the concentration of carbon dioxide in the calcining furnace is zero.

From the data below, calculate the following.

(a) The pressure of carbon dioxide at the surface of the limestone at 1000 K.
(b) The rate of decomposition of limestone at 1000 K (in moles per square meter of surface per second).
(c) Estimate the decomposition rate at 1100 K.
(d) Assume that at low temperatures (<900 K) the calcination reaction rate is controlled by the chemical decomposition of calcium carbonate. Sketch (in schematic form) the temperature dependence of the decomposition rate from 700 to 1100 K. Use appropriate functions for the scales on the two axes.

DATA

For the reaction $CaO + CO_2 \rightarrow CaCO_3$:
$$\Delta G° = -161{,}300 + 137T \text{ J/mol}$$
$$\text{using } P = 1 \text{ atm as the standard state}$$

Diffusion coefficient for CO_2 in air at 1000 K = 4.7×10^{-4} m^2/s

Chapter 8

Nonequilibrium Thermodynamics

Classical thermodynamics is concerned with systems in equilibrium. The concepts of energy and entropy are interpreted in the framework of equilibrium states, and changes in these quantities are evaluated as changes between equilibrium states. In fact, the power of classical thermodynamics derives from the definitions of energy and entropy, which make them point or state functions, independent of the path by which a system arrived at the particular state. For example, the change in entropy or energy of a specific quantity of gas is determined by temperatures and pressures of that gas, at the beginning and end of the process for which the change is being calculated, independent of the pressure–temperature path followed by the gas during the process. Thus, we think of changes in energy or entropy as the change in these quantities between equilibrium states.

In many situations of interest in materials science and engineering, however, the materials in question do not exist at equilibrium: either they are changing in some way, or they exist in a steady state in which some intensive properties are not equal throughout the system. Processes such as diffusion and chemical reactions are examples of the former. A material existing in a steady (time-invariant) temperature gradient is an example of the latter. In a strict sense, classical thermodynamics should not be applied to these situations. Using temperature as an example, it is philosophically difficult to define temperature at a point in a bar of material with a temperature gradient in it. When the bar comes to equilibrium, its temperature will be the same everywhere, and at that point we can speak of the temperature of the system and calculate the thermodynamic properties of interest. But we have no reason to believe that our calculations would be valid at some particular point in a bar whose ends were at different temperatures.

We can make the leap from classical thermodynamics to thermodynamics in nonequilibrium systems if we hypothesize that thermodynamics is valid for systems not too far displaced from equilibrium. In the case of the bar with a temperature gradient, we could imagine a very thin slice of the bar to be isolated long enough for the temperature within that infinitesimal slice to become equilibrated, and we could think of that as the temperature in the center of the slice. If we can make that hypothesis, we may apply the thermodynamic formulas to nonequilibrium systems.

The study of nonequilibrium thermodynamics (sometimes called irreversible thermodynamics) seeks to establish a framework for dealing with systems in which some flow is occurring. As a beginning, it is interesting to note the parallelism in several phenomenological laws governing flow: the flow of matter (Fick's law), the flow of heat (Fourier's law), and the flow of electrical charge (Ohm's law). These laws can be written, for one-dimensional flow, as follows:

$$J_M = -D \frac{dC}{dx} \tag{8.1}$$

$$J_Q = -k \frac{dT}{dx} \tag{8.2}$$

$$J_E = -\frac{1}{\rho} \frac{d\Phi}{dx} \tag{8.3}$$

In these, the flux of matter (J_M) is related to the diffusivity and the gradient of the concentration. The flux of heat (J_Q) is related to the thermal conductivity and the gradient of temperature. The flux of electrical charge (i.e., the current, J_E) is related to the resistivity and voltage gradient or the electric field.

We can think of these equations in general terms as

$$J = LX \tag{8.4}$$

For each force X, there is a corresponding conjugate primary flow, J (Eqs. 8.1–8.3). The question arises, Can there be cross-terms? Can a thermal gradient (dT/dx) influence the flow of electrical charge or the flow of mass? In both cases, the answer is yes. The first is illustrated by a thermocouple in which an electrical potential difference is generated in a wire because of a difference in temperature between the ends. In the second, the phenomenon is called thermal diffusion. Chemical potential differences or concentration differences can be generated by temperature gradients. These phemonena, and others like them, are called cross-effects.

If there are valid cross-effects, then the flux–force relationship (Eq. 8.4) should really be written as a matrix:

$$J_i = \sum_k L_{ik} X_k \tag{8.5}$$

The diagonal terms in this matrix ($i = k$) represent the phenomenological equations of Fick, Fourier, and Ohm. The off-diagonal terms ($i \neq k$) represent other effects, usually lower in physical magnitude, but nonetheless of some importance.

8.1 ENTROPY GENERATION

The linkage among the phenomena mentioned above is the concept of entropy generation. When forces and fluxes are defined in a way that relates to entropy generation, some interesting relationships among the off-diagonal coefficients (L_{ik}) can be developed. For example, these relationships link some physical phenomena, such as the thermocouple effect (Seebeck) and Peltier heating or cooling in junctions between materials.

To illustrate entropy generation in the case of thermal energy flow (heat flow), consider two chambers, I and II, at different temperatures, separated by a thermally conductive partition (see Figure 8.1).

If a quantity of heat ∂Q flows from I to II, the change in entropy in the two reservoirs is:

$$dS_{\mathrm{I}} = - \frac{\delta Q}{T_{\mathrm{I}}}$$

$$dS_{\mathrm{II}} = + \frac{\delta Q}{T_{\mathrm{II}}}$$

The net change in entropy is equal to

$$dS_{\mathrm{net}} = dS_{\mathrm{I}} + dS_{\mathrm{II}} = \delta Q \left(\frac{1}{T_{\mathrm{II}}} - \frac{1}{T_{\mathrm{I}}} \right) = \delta Q \left(\frac{T_{\mathrm{I}} - T_{\mathrm{II}}}{T_{\mathrm{I}} T_{\mathrm{II}}} \right) \tag{8.6}$$

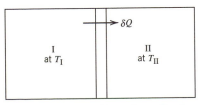

Figure 8.1 Flow of heat from chamber I to chamber II.

Entropy is not conserved. The entropy generated by the process of heat flow is given by Eq. 8.6.

Let us define a term θ, which is equal to the rate of entropy generation per unit volume.

$$\theta = \frac{1}{V}\frac{dS}{dt} \tag{8.7}$$

Let us now evaluate θ in the partition of Figure 8.1. Think of the partition as having infinitesimal thickness, dx. In this case,

$$dS = \delta Q\,\frac{d(1/T)}{dx}\,dx = \delta Q\,\text{grad}\left(\frac{1}{T}\right)dx$$

$$d\dot{S} = \frac{dS}{dt} = \frac{\delta Q}{dt}\,\text{grad}\left(\frac{1}{T}\right)dx$$

$$\theta = \frac{d\dot{S}}{V} = \frac{d\dot{S}}{A\,dx} = J_Q\,\text{grad}\left(\frac{1}{T}\right) \tag{8.8}$$

where $J_Q = \delta Q/A\,dt$.

The entropy generation term, θ, can be seen to be the product of the flux and a corresponding, or conjugate, driving force. The flux in this case is the flux of heat. The driving force is the gradient of the reciprocal of temperature. We will derive a set of conjugate fluxes and forces for other phenomena.

In one of the seminal works on the subject of nonequilibrium thermodynamics, Onsager defined the conjugate forces and fluxes so that their product is the temperature multiplied by the entropy generation term,[1] $T\theta$ (see Refs. 1 and 2). In these terms, the heat transfer during the nonequilibrium processes described earlier can be treated as follows:

$$T\theta = J_Q T\,\text{grad}\left(\frac{1}{T}\right)$$

$$T\theta = J_Q T\left(-\frac{1}{T^2}\frac{dT}{dx}\right)$$

$$T\theta = J_Q\left(-\frac{1}{T}\frac{dT}{dx}\right) \tag{8.9}$$

---

[1]The term $T\theta$, the entropy generation rate per unit volume multiplied by the temperature, is equivalent to an energy dissipation rate per unit volume.

Thus in heat transfer the flux, J_Q, is the flow of heat per unit area per unit time. The conjugate force is

$$X_Q = -\frac{1}{T}\frac{dT}{dx} \tag{8.10}$$

One of the terms in the matrix described earlier (Eq. 8.5) relates the flow of heat to the conjugate thermal driving force,

$$J_Q = L_{QQ}X_Q$$

From Fourier's law,

$$J_Q = -k\frac{dT}{dx}$$

Thus the coefficient in the matrix of Eq. 8.5 is equal to the temperature times the thermal conductivity[2]:

$$L_{QQ} = Tk \tag{8.11}$$

8.2   FORCES AND FLUXES

To incorporate other flows, such as mass and electrical charge, into the matrix of Eq. 8.5, we begin by writing the property relationship for the change of internal energy of a system (Eqs. 1.21 and 1.22).

$$dU = T\,dS - P\,dV + \Sigma\mu_i dn_i + \Phi\,dQ \tag{8.12a}$$

$$dS = \frac{dU}{T} + \frac{P}{T}\,dV - \sum_i \frac{\mu_i}{T}\,dn_i - \frac{\Phi}{T}\,dQ \tag{8.12b}$$

To develop the form of forces and fluxes in mass flow, treat the case of no P-V work, no heat flow, and no flow of charge. When these conditions are

---

[2]The same problem may be handled by considering the product of forces and fluxes to be simply the entropy generation term. In this case,

$$L_{QQ} = T^2k$$

Both methods for treating forces and fluxes are valid. The values of the coefficients in the matrix will be different by a multiple of the temperature, T. The conclusions drawn by comparing the various terms in the matrices will be the same, however, no matter which method is used.

applied to Eq. 8.12b, we obtain:

$$\theta = -\sum_i J_i \, \text{grad} \left(\frac{\mu_i}{T}\right) \tag{8.13}$$

Using a similar technique, for the flow of electrical charge one obtains

$$\theta = -i \, \text{grad} \left(\frac{\Phi}{T}\right) \tag{8.14}$$

For the flow of heat, mass, and charge, the product of temperature and entropy generation rate is:

$$T\theta = J_Q T \, \text{grad} \frac{1}{T} - T\sum_i J_i \, \text{grad} \left(\frac{\mu_i}{T}\right) - TJ_E \, \text{grad} \left(\frac{\Phi}{T}\right) \tag{8.15}$$

Corresponding to the flows of heat J_Q, flow of mass J_M, and flow of charge J_E, we have the following conjugate forces:

Thermal: $$X_Q = T \, \text{grad} \left(\frac{1}{T}\right) = -\frac{1}{T}\frac{dT}{dx} \tag{8.16}$$

Mass: $$X_{M_i} = -T \, \text{grad} \left(\frac{\mu_i}{T}\right) \tag{8.17}$$

Electrical: $$X_E = -T \, \text{grad} \left(\frac{\Phi}{T}\right) \tag{8.18}$$

8.3 ONSAGER RECIPROCITY RELATIONSHIP

In the matrix represented by Eq. 8.5,

$$J_i = \sum_k L_{ik} X_k \tag{8.5}$$

the coefficients L_{ik} relating the conjugate forces and fluxes (L_{ii}) have been discussed. The off-diagonal terms ($i \neq k$) are also of interest.

A major contribution to the study of nonequilibrium thermodynamics was made by Onsager, who showed that using a principle of microscopic reversibility (Refs. 1 and 2), one could demonstrate that

$$L_{ik} = L_{ki} \tag{8.19}$$

This relationship (Eq. 8.19) applies to the form of the matrix shown in Eq. 8.5 and also in the case of some transformations of that matrix that are used to relate flows to more commonly observed gradients, such as chemical potential gradients or the negative gradient of electrical potential, the electric field.

8.4 THERMAL DIFFUSION

The presence of a thermal gradient in a system can cause the flow of a single component relative to its container. In homogeneous, multicomponent solutions, the presence of a temperature gradient can cause relative movements of the components, that is, an unmixing. This phenomenon is called *thermal diffusion,* and also the *Soret effect.* We discussed an example of thermal diffusion in Chapter 2 (Section 2.12), namely, a gas held in two chambers connected by a very fine hole.[3] If the chambers are at different temperatures, T_1 and T_2, a pressure difference will be generated between them. When the two chambers come to steady state, where no net flow occurs between them, the pressures in the two chambers will be:

$$\frac{P_1}{P_2} = \left(\frac{T_1}{T_2}\right)^{1/2}$$

(2.34)

This is an example of thermal diffusion, or *thermal transpiration,* as it is sometimes called when gases are involved.

In terms of the equations we derived earlier, thermal diffusion can be represented by the terms in the matrix (Eq. 8.5) that pertain to the temperature gradients and gradients of chemical potential as follows:

$$J_Q = -L_{QQ}\frac{1}{T}\frac{dT}{dx} - L_{QM}T\frac{d}{dx}\left(\frac{\mu}{T}\right)$$

(8.20)

$$J_M = -L_{MQ}\frac{1}{T}\frac{dT}{dx} - L_{MM}T\frac{d}{dx}\left(\frac{\mu}{T}\right)$$

(8.21)

When a material is held in a temperature gradient until it comes to steady state, the mass flow J_M will be zero and we have, from Eq. 8.21:

$$L_{MQ}\frac{1}{T}\frac{dT}{dx} = -L_{MM}T\frac{d}{dx}\left(\frac{\mu}{T}\right)$$

(8.22)

Eliminating dx,

$$d\left(\frac{\mu}{T}\right) = -\frac{L_{MQ}}{L_{MM}}\frac{dT}{T^2}$$

---

[3]The size of the hole is determined by the mean free path of the gases in the two chambers. For this analysis to be valid, the hole must be small compared to the mean free path; that is, the flow between the two must be by molecular flow (Knudsen flow), not viscous flow.

Differentiating,

$$d\left(\frac{\mu}{T}\right) = \frac{1}{T} d\mu + \mu\, d\left(\frac{1}{T}\right)$$

From thermodynamic relations,

$$d\mu = -\overline{S}\, dT + \overline{V}\, dP \qquad \text{and} \qquad \mu = \overline{H} - T\overline{S}$$

Hence:

$$d\left(\frac{\mu}{T}\right) = \frac{\overline{V}\, dP}{T} - \overline{H}\,\frac{dT}{T^2} = -\frac{L_{MQ}}{L_{MM}}\frac{dT}{T^2} \tag{8.23}$$

$$\frac{\overline{V}\, dP}{T} = \left(\overline{H} - \frac{L_{MQ}}{L_{MM}}\right)\frac{dT}{T^2} \tag{8.24}$$

To gain an understanding of the term L_{MQ}/L_{MM}, consider the ratio of heat (energy) flow to mass flow under isothermal conditions, that is, where $dT/dx = 0$. Dividing Eqs. 8.20 and 8.21,

$$\frac{J_Q}{J_M} = \frac{L_{QM}}{L_{MM}} \tag{8.25}$$

By the Onsager relation (Eq. 8.19), $L_{QM} = L_{MQ}$, hence:

$$\frac{J_Q}{J_M} = \frac{L_{MQ}}{L_{MM}} \tag{8.26}$$

Based on Eq. 8.26, the term L_{MQ}/L_{MM} represents the energy flow associated with a mass flow. Defining $Q' \equiv L_{MQ}/L_{MM}$, we write

$$\frac{\overline{V}}{T}\, dP = (\overline{H} - Q')\frac{dT}{T^2} \tag{8.27}$$

Let us define

$$Q^* \equiv Q' - H \tag{8.28}$$

where Q^* represents the difference between the energy *associated with* the material that flows, Q', and the enthalpy (partial molar enthalpy) of the material in the reservoirs *from which* it flows.

$$\frac{\overline{V}}{T}\, dP = -Q^* \frac{dT}{T^2} \tag{8.29}$$

With this as background, let us return to the thermal transpiration situation. It can be shown (Appendix 8A) that the energy of monatomic gas molecules moving between the two chambers by Knudsen flow is $2RT$ joules per mole. The energy of the gas in a chamber is $\frac{3}{2}RT$ joules per mole. The enthalpy is $U + PV$ or $U + RT$ for an ideal gas. The enthalpy is thus $\frac{5}{2}RT$ joules per mole. The term Q^* is thus $-\frac{1}{2}RT$. Substituting this value in Eq. 8.29, and recognizing that $V/T = R/P$, we have

$$\frac{dP}{P} = \frac{1}{2}\frac{dT}{T}$$

Integrating the preceding equation gives:

$$\int_{P_2}^{P_1} \frac{dP}{P} = \frac{1}{2}\int_{T_2}^{T_1} \frac{dT}{T}$$

$$\frac{P_1}{P_2} = \left(\frac{T_1}{T_2}\right)^{1/2}$$

The formalism of nonequilibrium thermodynamics yields the same result (Eq. 2.34) as the Langmuir equation in the case of thermal transpiration of a gas.

In a homogeneous solution, the individual components react differently in a temperature gradient. One component diffuses preferentially up the gradient, which enriches the hot end with that component. The relationship between concentration and temperature at steady state can be derived by setting the term $V\,dP$ in Eq. 8.29 for component 1 equal to $RT\,d(\ln a_1)$. If the thermodynamic activity of component 1 is linearly proportional to its concentration, then $V\,dP = RT\,d(\ln C_1)$. The relationship between the concentration gradient and the temperature gradient at steady state is then:

$$\frac{d\ln C_1}{dx} = -\frac{Q_1^*}{RT^2}\frac{dT}{dx} \tag{8.30}$$

Unmixing of homogeneous alloys can take place when a material is held in a temperature gradient at temperatures high enough to allow for atomic mobility. Shewmon in 1960 (Ref. 3) demonstrated a thermal diffusion effect in iron–carbon alloys. In his experiments he subjected samples of iron–carbon alloys to temperature gradients until steady state concentration gradients were established. In iron–carbon alloys, carbon moves preferentially to the hotter side. Shewmon determined the value of Q^* for α-iron to be 24 ± 1.5 kcal/mol at 700°C. The value for γ-iron was much lower. Other examples of thermal diffusion can be found in materials subjected to temperature gradients at high temperatures, such as the cladding material used in nuclear reactor fuel elements, and in oxide nuclear fuel pellets themselves.

8.5   ELECTROMIGRATION

Just as mass can be moved by a temperature gradient, so mass can be moved by an electrical potential gradient. This phenomenon is called *electromigration*. The flux equation for mass flow is:

$$J_M = -L_{MM}T \text{ grad} \left(\frac{\mu}{T}\right) - L_{ME}T \text{ grad} \left(\frac{\Phi}{T}\right) \tag{8.31}$$

At constant temperature, this becomes:

$$J_M = -L_{MM} \text{ grad } \mu - L_{ME} \text{ grad } \Phi \tag{8.32}$$

To identify some of the coefficients in Eq. 8.32, consider the mass flow in one direction (x) with no electric potential gradient present:

$$J_M = -L_{MM} \text{ grad } \mu = -L_{MM}RT \frac{d \ln a}{dx}$$

For a solution in which thermodynamic activity a is linearly proportional to concentration, which we will designate as C:

$$J_M = -\frac{L_{MM}RT}{C} \frac{dC}{dx} = -D \frac{dC}{dx} \tag{8.33}$$

Hence,

$$L_{MM} = \frac{DC}{RT} \tag{8.34}$$

At constant temperature, with no chemical potential gradient:

$$J_M = -L_{ME} \text{ grad } \Phi = L_{ME}E$$

where E is the electric field.

The mass flow rate can be equated to the concentration of the species in question multiplied by a drift velocity v.

$$J_M = Cv$$

The drift velocity is the mobility of the species, B, multiplied by the force on it, F, which in this case is Z^*eE:

$$v = BF = BZ^*eE \tag{8.35}$$

where Z^* is the effective valence of the species, e is the charge of the electron, and E is the electric field.

In terms of the diffusion coefficient (for an ideal solution), the mobility is:

$$B = \frac{D}{kT}$$

The mass flux can then be written as follows:

$$J_M = \frac{DC}{kT} Z^*eE \tag{8.36a}$$

This can also be written

$$J_M = \frac{DC}{RT} Z^*\mathscr{F} E \tag{8.36b}$$

where $\mathscr{F}$ is the Faraday constant

The mass flow equation with a concentration gradient and an electric field present is:

$$J_M = \frac{DC}{RT} \left(-\frac{RT \, d(\ln C)}{dx} + Z^*\mathscr{F} E \right) \tag{8.37a}$$

This can also be written

$$J_M = -D \frac{dC}{dx} + \frac{D}{RT} CZ^*\mathscr{F} E \tag{8.37b}$$

From Eqs. 8.37, we see that a material can move in response to a concentration gradient, and also in response to an electric field. The magnitude of the effect depends on the magnitude of the diffusion coefficient, and therefore the temperature at which the phenomenon is observed. It also depends on the magnitude of Z^*, the effective valence of the species in question. Metallic elements tend to have negative values of Z^*. Gold, for example, has a Z^* of about -8. This is strange because our view of a metal consists of positive ions of the element among free electrons. The explanation given for the negative Z^* is the "electron wind" effect: that is, the metallic atoms interact with the electrons moving under the influence of the electric field, and are dragged along by them.

At steady state, when the net flow of matter J_M is zero:

$$\frac{1}{C}\frac{dC}{dx} = \frac{d(\ln C)}{dx} = \frac{Z^*\mathscr{F} E}{RT} \tag{8.38}$$

Equation 8.38 shows that a concentration gradient will be established by an electric field at steady state in multicomponent solutions. The movement of material under the influence of an electric field has also been observed in single-component systems where current densities are high. The movement of tungsten in the filaments of direct-current incandescent lamps was observed many years ago. A more modern example involves metal lines in integrated circuits. Although the currents in these circuits are low, on the order of milliamperes, the current densities are high. A current of one milliampere through a line with cross-sectional area of 10^{-8} cm^2 means a current density of 10^5 A/cm^2. At this level of current density, materials seem to move. In fact, distortions of lines produced by such movement are responsible for failures in some integrated circuits.

8.6   THERMAL–ELECTRICAL EFFECTS

Two of the terms in the flow–flux matrix (Eq. 8.5) relate thermal effects to electrical effects in a material. Unless the coefficients L_{QE} and L_{EQ} are zero, temperature gradients should be expected to generate electrical potential gradients. In addition, heat flow should be influenced not only by the temperature gradient, but also by the electrical potential gradient. There are three such thermal–electrical effects: the Seebeck effect, the Peltier effect, and the Thomson effect.

The Seebeck effect is the basis of the thermocouple, a device commonly used to measure temperature differences. In a thermocouple two different materials, usually in the form of wires, are joined as shown in Figure 8.2. If the junctions BA and AB are at different temperatures, an electric potential difference (voltage) will be observed between points 1 and 2. That voltage, $\Delta\Phi$, measured by a potentiometer in the absence of current flow ($J_E = 0$), is related to the temperature difference between the two junctions, ΔT. The Seebeck coefficient ε_{AB} is a function of the two

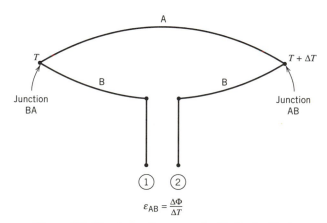

Figure 8.2 Circuit for thermocouple (Seebeck effect).

materials, A and B. From Eq. 8.39, it can be seen that ε_{AB} relates the observed voltage $\Delta\Phi$ to the temperature differences between the two junctions ΔT.

$$\varepsilon_{AB} = \left(\frac{\Delta\Phi}{\Delta T}\right)_{J_E=0} \tag{8.39}$$

The magnitude of the Seebeck coefficient is on the order of 15 μV/K for common combinations of metals. Semiconductors can have values of many hundreds of microvolts per kelvin. It should be emphasized that ε_{AB} is a property of the *combination* of materials A and B. However, one finds tabulations of the thermoelectric power of *individual* materials. In these tabulations the Seebeck coefficients are given for the combination of the individual material and pure lead.[4] Sometimes ε_{AB} is expressed in the form $a + bT$, where T is the temperature. To determine the Seebeck coefficient (ε_{AB}) for a combination of materials, the values of a and b for one material in the combination are subtracted from the values for the other. For example, the ε_{AB} for the copper–iron thermocouple may be calculated as follows:

For Cu: $\varepsilon_{Cu} = 1.34 + 0.0094T$ (μV/°C)
For Fe: $\varepsilon_{Fe} = 17.15 - 0.0482T$ (μV/°C)

$$\varepsilon_{Cu-Fe} = -15.81 + 0.0576 \ (\mu V/°C)$$

$$\varepsilon_{Cu-Fe} = \frac{d(\Delta\Phi_{Cu-Fe})}{dT}$$

A measurement of the voltage between points 1 and 2 in Figure 8.2 (with no current flowing) establishes the temperature difference between junctions AB and BA. Usually one of the junctions is kept in an ice-water bath (0°C). The voltage $\Delta\Phi$ is then a measure of the temperature of the other junction in degrees centigrade. For example, let us calculate the voltage that is to be expected in a copper–iron thermocouple when one junction is in an ice-water bath at 0°C and the other is in boiling water at 100°C.

$$\Delta\Phi = \int_0^{100} (-15.81 + 0.0576T)dT \ (\mu V)$$

$$\Delta\Phi = \left(-15.81T + \frac{0.0576}{2}T^2\right)\Bigg|_0^{100}$$

$$\Delta\Phi = -1293 \ \mu V$$

---

[4]Lead is chosen as the reference material because pure lead has very low thermoelectric power. In particular, it shows a very small Thomson effect (see the discussion following Eq. 8.57).

The relation between voltage and temperature difference determined by using thermoelectric power values in the form $a + bT$ is a useful approximation. For more accurate temperature determinations, tables of observed voltage as a function of temperature above the ice point ($T = 0°C$) are available for most commonly used thermocouples. In careful experimental work, thermocouples are usually calibrated.

Let us now consider a different thermal–electrical effect. Let the conditions shown in Figure 8.2 be modified so that the two junctions, AB and BA, are at the same temperature. When a current is passed through the wires, heat will be absorbed at one of the junctions and released at the other. This is the Peltier effect (Figure 8.3). The Peltier coefficient is defined as the heat absorption (or release) at a junction per unit charge flow. It may also be expressed in terms of rates:

$$\Pi_{T,AB} = \frac{J_{Q,AB}}{J_E} \tag{8.40}$$

The subscripts T and AB on the Peltier coefficient remind us that it is a function of temperature as well as of the combination of two materials at the junction, A and B. The effect is reversible. If the direction of current flow is reversed, the electrode that absorbed heat before the reversal will now release heat.

The Peltier heat at a junction, absorbed or released, is in addition to the Joule heat (i^2R) that is generated in the junction. It is possible to generate a net cooling effect at a junction; that is, the Peltier cooling can be greater than the Joule heating. Peltier junctions are very useful for accurately controlling temperatures of small samples because electrical currents can be well controlled and monitored.

The relationship between the Peltier coefficient and the Seebeck coefficient for a combination of materials can be established using the matrix expression for flow and fluxes (Eq. 8.5) and the Onsager relation (Eq. 8.19). If the matrix of Eq. 8.5 is reduced to the terms involving electrical and thermal terms, we have

$$J_Q = L_{QQ}T \text{ grad}\left(\frac{1}{T}\right) - L_{QE}T \text{ grad}\left(\frac{\Phi}{T}\right)$$

$$J_E = L_{EQ}T \text{ grad}\left(\frac{1}{T}\right) - L_{EE}T \text{ grad}\left(\frac{\Phi}{T}\right) \tag{8.41}$$

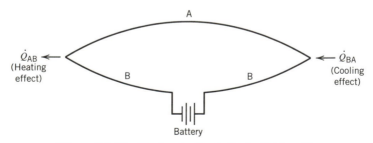

Figure 8.3 Thermoelectric circuit for the Peltier effect.

These equations can be transformed to the more usual driving forces, dT/dx and $d\phi/dx$, as follows:

$$J_Q = -L_{QQ} \frac{1}{T} \frac{dT}{dx} - L_{QE} T \left[\frac{1}{T} \text{grad } \Phi - \Phi \frac{1}{T} \frac{dT}{dx} \right]$$

$$J_Q = \left(-L_{QQ} + \Phi L_{QE} \right) \frac{1}{T} \frac{dT}{dx} - L_{QE} \frac{d\Phi}{dx} \tag{8.42}$$

Similarly,

$$J_E = \left(-L_{EQ} + \Phi L_{EE} \right) \frac{1}{T} \frac{dT}{dx} - L_{EE} \frac{d\Phi}{dx}$$

If we now define new coefficients as follows:

$$L_{QQ'} = L_{QQ} - L_{QE}\Phi \quad \text{and} \quad L_{EQ'} = L_{EQ} - L_{EE}\Phi \tag{8.43}$$

the transformed matrix is:

$$J_{Q'} = -L_{QQ'} \frac{1}{T} \frac{dT}{dx} - L_{QE} \frac{d\Phi}{dx} \tag{8.44}$$

$$J_{E'} = -L_{EQ'} \frac{1}{T} \frac{dT}{dx} - L_{EE} \frac{d\Phi}{dx} \tag{8.45}$$

This matrix has the advantage of using driving forces that are easier to conceptualize: the temperature gradient and the gradient of electrical potential.

The Peltier coefficient was defined as the ratio of heat flow to current flow in a junction in the absence of temperature differences, $dT/dx = 0$. When this condition is applied to Eq. 8.44 and 8.45, they become

$$J_{Q,AB} = -L_{QE} \frac{d\Phi}{dx} \tag{8.46}$$

$$J_{E,AB} = -L_{EE} \frac{d\Phi}{dx} \tag{8.47}$$

Dividing Eqs. 8.46 and 8.47 yields:

$$\frac{J_{Q,AB}}{J_{E,AB}} = \frac{L_{QE}}{L_{EE}} = \Pi_{T,AB} \tag{8.48}$$

The Seebeck coefficient was defined at zero current flow, $J_E = 0$. Under this condition, Eq. 8.45 becomes

$$0 = -L_{EQ'} \frac{1}{T} \frac{dT}{dx} - L_{EE} \frac{d\Phi}{dx} \tag{8.49}$$

Then:

$$\frac{d\Phi}{dT} = -\frac{1}{T} \frac{L_{EQ'}}{L_{EE}} \tag{8.50}$$

From the Onsager relationship, $L_{QE} = L_{EQ'}$, hence:

$$\frac{d\Phi}{dT} = -\frac{1}{T} \frac{L_{QE}}{L_{EE}} \tag{8.51}$$

Combining Eqs. 8.51, 8.48, and 8.39 establishes the relationship between the Peltier and Seebeck coefficients:

$$\left(\frac{\Delta\Phi}{\Delta T} \right)_{J_E=0} = -\frac{\Pi_{T,AB}}{T} = \varepsilon_{AB} \tag{8.52}$$

The third thermal–electrical phenomenon to be discussed is the Thomson effect. Unlike the Peltier and Seebeck effects, which apply to combinations of two materials, the Thomson effect is concerned with potential differences, current flows, and temperature gradients in a single material. The wire in Figure 8.4 is in a temperature gradient. If no current is flowing in the wire, there will be a difference in potential

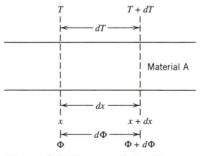

Figure 8.4 Diagram of the Thomson effect.

between the points x and $x + dx$. This is one way of looking at the Thomson effect. In these terms, the Thomson coefficient of the material A, σ_A, is the ratio of the potential difference to the temperature difference:

$$\sigma_A = \frac{d\Phi}{dT} \tag{8.53}$$

It can be shown that the statement of Eq. 8.53 is equivalent to saying that when an electric current is passed through the wire (in a temperature gradient), there will be some heat released or absorbed in the wire at a rate in excess of the Joule heat (i^2R or $J_E^2 R$). This excess heat is the Thomson heat. The effect is reversible; that is, the excess heat that is released if the current flow is in one direction will be absorbed if the current direction is reversed. The ratio of Thomson heat to charge passed through the wire is related to the Thomson coefficient:

$$\frac{J_Q}{J_E} = \sigma_A dT \tag{8.54}$$

To derive the relationship between the Thomson, Peltier, and Seebeck coefficients, consider the thermocouple circuit of Figure 8.2. Let a unit charge be passed through the circuit. An energy balance around the circuit is:

$$\Pi_{T,BA} + \Pi_{T+dT,AB} + (\sigma_B - \sigma_A)dT = d\Phi \tag{8.55}$$

$$\Pi_{T,BA} - \Pi_{T+dT,BA} + (\sigma_B - \sigma_A)dT = d\Phi$$

The term $\Pi_{T+dT,BA}$ can be expressed as follows:

$$\Pi_{T+dT,BA} = \Pi_{T,BA} + \frac{d\,\Pi_{T,BA}}{dT}\,dT = \Pi_{T,BA} + d\,\Pi_{T,BA}$$

Then:

$$-d\,\Pi_{T,BA} + (\sigma_B - \sigma_A)dT = d\Phi \tag{8.56}$$

From $\Pi_{T,AB} = -T\varepsilon_{AB}$, $d\,\Pi_{T,AB} = -T\,d\varepsilon_{AB} - \varepsilon_{AB}\,dT$. Substituting in Eq. 8.56 and dividing by dT, we have

$$T\frac{d\varepsilon_{AB}}{dT} + \varepsilon_{AB} + (\sigma_B - \sigma_A) = \frac{d\Phi}{dT} = \varepsilon_{AB}$$

Finally:

$$\sigma_A - \sigma_B = T \frac{d\varepsilon_{AB}}{dT} \tag{8.57}$$

The Thomson coefficients σ_A and σ_B are related to the temperature derivative of the Seebeck coefficient. Independent Thomson coefficients can be evaluated by using pure lead as one of the materials, because it has been found that pure lead has a zero Thomson coefficient.

REFERENCES

1. Denbigh, K. G., *Thermodynamics of the Steady State,* Methuen, London, 1951.
2. Prigogine, I., *Introduction to Thermodynamics of Irreversible Processes,* Charles C. Thomas, Springfield, IL, 1955.
3. Shewmon, P., *Acta Met., 8,* 606 (1960).

APPENDIX 8A

Average Energy of Effusing Molecules

This appendix contains the derivation of the average energy of gas molecules effusing (by molecular flow) from a hole in a container.

From Vol. I, Chapter 10 (Section 10.1, Eqs. 10.5–10.8 and Section 10.11, Eq. 10.62), the average translational energy of a gas molecule is $E = \frac{3}{2}kT$. This is the result of motion in three directions: x, y, and z. In each direction the average energy is $\frac{1}{2}kT$.

When a gas effuses through a hole in a container in the x direction, the average energy of the effusing molecules in the y and z directions is the same as the average energy in the y and z directions in the container. But the effusing molecules in the x direction have a different average energy. From Section 2.12 (Eq. 2.26a and subsequent integral form, Eq. 2.26b), the number of molecules flowing from the hole can be represented by

$$N^* = K \int_0^\infty v_x \exp\left(-\frac{mv_x^2}{2kT}\right) dv_x \tag{8A.1}$$

where K is a constant at any given temperature.

The average squared velocity of the molecules escaping in the x direction is

$$\overline{v_x^2} = \frac{\displaystyle\int_0^\infty v_x^3 \exp\left(-\frac{mv_x^2}{2kT}\right) dv_x}{\displaystyle\int_0^\infty v_x \exp\left(-\frac{mv_x^2}{2kT}\right) dv_x} = \frac{\dfrac{1}{2}\left(\dfrac{2kT}{m}\right)^2}{\dfrac{1}{2}\left(\dfrac{2kT}{m}\right)} = \frac{2kT}{m} \qquad \text{(8A.2)}$$

The average energy of the effusing molecules in the x direction is then:

$$E_x = \tfrac{1}{2}m\overline{v_x^2} = kT$$

The total energy of the effusing molecules is:

$$E = E_x + E_y + E_z = kT + \tfrac{1}{2}kT + \tfrac{1}{2}kT = 2kT$$

Index